THE
DEMING
DIMENSION

Other books published by SPC Press:

The World of W. Edwards Deming, Second Edition
by
Cecelia Kilian

Deming's Route to Continual Improvement
by
William W. Scherkenbach

Real People, Real Work
by
Lee Cheaney and Maury Cotter

SPC at the Esquire Club
by
Donald J. Wheeler

Understanding Statistical Process Control, Second Edition
by
Donald J. Wheeler and David S. Chambers

THE
DEMING
DIMENSION

HENRY R. NEAVE

SPC Press

Knoxville, Tennessee

COPYRIGHT © 1990 HENRY R. NEAVE

SPC Press, Inc.
5908 Toole Drive, Suite C
Knoxville, Tennessee 37919   USA
Telephone   (615) 584-5005
Fax   (615) 588-9440

ISBN 0-945320-08-6

2345678910

CONTENTS

PART 2:
SOME FUNDAMENTALS

PART 3:
THE NEW CLIMATE

PART 4:
FOUNDATIONS OF KNOWLEDGE

PART 5:
THE 14 POINTS REVISITED

FOREWORD
BY
W. EDWARDS DEMING

Western management in industry, education, government, is due for sweeping changes. The prevailing system of management has smothered the individual, and has consequently dampened innovation, applied science, joy in learning, joy in work. It will be necessary to restore dignity and self-esteem to the individual. This can be done, but only by transformation of the style of management now practised.

The prevailing forces of destruction start early in life—grades in school from toddler on up through the university, gold stars for school athletics, merit system or annual appraisal on the job, incentive pay, work standards, MBO (rather, MBIR: Management by Imposition of Results), MBR (Management by Results). These forces of destruction must be replaced by leadership.

Leadership requires awareness and respect for differences between people. It is necessary to adapt these different personalities to optimisation of the work of the group.

The prevailing style of management was not born with evil intent. It grew up little by little by reactive behaviour, unsuited to any world, and especially unsuited to the new kind of world of dependence and interdependence that we are in now.

The cost of the prevailing system of management is failure to keep up with quality, loss of competitive position, and destruction of the individual.

It will be necessary to learn and to practise a System of Profound Knowledge.

A System of Profound Knowledge consists of four parts:

> Knowledge about a system;
> Some knowledge about variation;
> Some theory of knowledge;
> Some psychology.

The various segments of Profound Knowledge cannot be separated. They interact with one another. Thus, knowledge of psychology is incomplete without knowledge of variation. If psychologists understood variation, they could no longer participate in continual refinements of instruments for rating people on a job.

The theory of knowledge helps us to understand that management in any form is prediction: that grades given in school are predictions of performance: that appraisal of employees is prediction.

Statistical theory can play a vital part in optimisation of a system. Statistical theory is helpful for understanding differences between people, and interactions between people and the system that they work in, or learn in.

A system is a series of functions or activities (hereafter components) within an organisation that work together toward the aim of the organisation. There is in almost any system interdependence between the components thereof.

The aim of the system must be stated by the management thereof. Without an aim, there is no system. The components of a system must be managed.

The performance of any component is to be evaluated in terms of its contribution to the aim of the system, not for its individual production or profit, nor for any other competitive measure.

Best efforts and hard work will not suffice, nor new machinery, computers, automation, gadgets. One could well add that we are being ruined by best efforts put forth with the best intentions but without guidance of a theory of management for optimisation of a system. There is no substitute for knowledge.

The change required is transformation, change of state, metamorphosis. The transformation will restore the individual; will abolish grades in school on up through the university; will abolish the annual appraisal of people on the job, MBO, quotas for production, specified requirements that people work 57 minutes out of every hour, incentive pay, monthly or quarterly reports on business targets, competition between people, competition between divisions, and other forms of suboptimisation. Leadership will replace these bad practices, and will restore the individual.

The transformation must be led by top management. The transformation is not stamping out fires, solving problems, nor cosmetic improvements.

This book by Henry Neave explains why the prevailing system of management has led us into decline. It explains the transformation that must for survival take place under the leadership of top management.

It gives me great pleasure and satisfaction to see this book, and to express my appreciation for the efforts of Dr. Neave to achieve clarity in exposition that carries the message to the management of industry, of education, and to people in government.

Washington
27 February 1990

W. Edwards Deming

AUTHOR'S FOREWORD

"Suppose," as Dr. Deming says in many of his presentations, "we were to hold a national referendum, and ask the question:

Are you in favour of quality?

Yes_____ No_____ "

Not unnaturally, he believes that the result would be an avalanche in favour of quality. There's no problem about that. The problem is that everybody has the answer on how to achieve quality. It seems so simple. Here are some of the answers that he suggests may be offered:

> Automation
> Computers
> Gadgets
> New machinery
> Make everybody accountable
> Posters, slogans, exhortations
> Zero Defects
> Management by Objective
> Incentive plans, bonuses
> Management by Results
> Work (time) standards, quotas
> Merit system, pay for performance
> Just-in-Time
> Meet specifications

Common sense
Good intentions
Best efforts
Hard work

Unfortunately, these answers are all wrong. Some are wrong in the sense that they are insufficient, while others are so wrong that they actually obstruct improvement of quality.

Other intended approaches to quality are based on some of the above answers. They are contained within a perception of management which is too limited. It is a perception which treats as *facts of life*, and therefore unchangeable, some of the most formidable obstacles to the Deming approach. So, while some accept those constraints and concentrate on working within them, one of Deming's principal concerns is to break free of them. And, once having broken free of them, most of the answers which look relevant from the limited view turn out to be irrelevant from the broader view.

Deming's work constitutes a new dimension of thought about management; hence the title of this book. The Deming philosophy teaches us the answers that we need, as opposed to those which inhibit and indeed reverse progress, so that our hard work and best efforts may be rewarded with success rather than disappointment.

PREFACE

Dr. Deming's seminal book *Out of the Crisis* is, I believe, the most important book that has ever been written *about* management, and the most important book that has ever been written *for* management. Sir John Egan, regarded by many as the saviour of Jaguar Cars during the 1980s, also regards it highly. In the September 1988 issue of *The Director*, Sir John wrote:

> "*Out of the Crisis* is required reading for every chief executive in British industry who is serious about ensuring the international competitiveness of his company."

But *Out of the Crisis* was published as long ago as 1986, around the time of Deming's 86th birthday. I have had the privilege to assist at a number of his four-day seminars since that time, and to be involved with several further seminars, conferences, and other meetings, both in Britain and America, in which Dr. Deming has participated. I am writing this Preface on his 89th birthday. During the intervening three years, Deming has developed his thinking by a staggering degree. Apart from anything else, that came home to me when I noticed the thickness of my piles of untidily-scribbled notes after each of his presentations. It is a matter of some wry amusement to hear of people saying that they have been to "the" Deming four-day seminar. *Which* four-day seminar? They're all different!

I originally set out with the intention to write a reasonably concise summary of developments in Deming's teachings since *Out of the Crisis*. Indeed, early and very abridged versions of what are now Parts 3 and 5 of this book are due to appear in the 1990 issues of a new journal: *Total Quality Management* (published in Sheffield, England).

However, I soon became dissatisfied on two counts with that very incomplete job. First, and not surprisingly, I found Deming saying many more things which were new to me at his presentations this year. Now, of course, for as long as he continues to draw breath (which, God willing, will be for a few more years yet), *any* book on Deming is bound to be out-of-date before it gets published; I know that. But that did mean I was less than happy with his suggestion to simply convert what I had already written for *Total Quality Management* into a book. And so I have now incorporated much material on what I have learned from him subsequently. Second, the papers in *Total Quality Management* had to assume the reader's knowledge of Deming's work as presented in *Out of the Crisis*. I was uncomfortable about that, but restrictions on space left no option.

I am concerned by the thought of people using those papers, in spite of my warnings, as *introductory* reading on the Deming philosophy. The Deming philosophy has been referred to by some as a "new religion"; I wouldn't necessarily go along with that, but the biblical analogy of the danger of building castles on sand does come to mind! Consequently, this book now begins with an attempt to summarise the philosophy as understood at the time *Out of the Crisis* was written. Such introductory work is certainly no replacement for *Out of the Crisis*, but I hope it will be helpful to those who have not yet read Dr. Deming's book. Indeed, I would like to think that one of the results of my writing this book is that it will become clear to those who are not familiar with *Out of the Crisis* how important it is that they become so. My frequent references to *Out of the Crisis* may serve as a kind of study-guide for that purpose.

One of the many reasons for needing familiarity with Dr. Deming's earlier work is to get a sense of the depth of vision embodied in his teachings. This is why I have also included a chapter (Chapter 2) sketching the history of his thankfully long life. Deming is a visionary—have no doubt of that, and do not scorn the fact. He had vision in 1950 when he first delivered lectures to the Japanese. He told them that (as long as they adopted his principles!) they would be able to take over important world markets. Yes, that was in 1950. Just think of the Japanese situation at that time—five years after the devastation of World War II, and with Japanese products suffering a well-deserved reputation of "cheap and shoddy." Even those top executives to whom Deming spoke in 1950 didn't really believe that what he said could be possible. But, in a sense, they had an advantage. They perceived the situation in Japan to be so bad that they surely didn't have anything to lose by learning from this extraordinary American statistician.

Well, that is past and proven history. There is much more contemporary proof of the accuracy and validity of Deming's visions. When the Americans eventually started listening to him around 1980, so much of what he then said seemed absurd —nice in theory, but not of this world. Ten years later, we would be wise indeed to consider how the world has changed, and how so much of that change has been in line with his predictions.

The perceived importance of quality itself is maybe the most obvious example. Who, in 1980, considered quality to be an important topic for the management agenda? It is now. Who entertained the notion of quality contributing positively rather than negatively to productivity and the reduction of costs? We do now. Who was concentrating then on the prevention of defects rather then their detection, and consequently on severe reduction of dependence on mass-inspection schemes? That change is now well-developed. Who was contemplating the single-supplier principle? Isn't it *obvious* that multiple-sourcing is astute business practice in order to keep everyone competitive? Not all readers will believe it, but thoughts on that have changed, and substantially.

In addition to my frequent references to *Out of the Crisis*, there is considerable direct input from Dr. Deming to this book. Chapter 1 and portions of Chapters 3 and 18 appear almost entirely in his own words, from both presentations and written material. Also many, many quotations from him are scattered throughout the book, and are highlighted as in the two examples below. I believe these are very valuable; it is quite some paradox that, whereas Deming regards his marathon four-day seminar as only a small beginning, he so often encompasses a depth of meaning in just a very few well-chosen words. As a taster, you might consider his description of both training for skills and improvement of processes, products, and services as "essential but unimportant."

Further, any delegates at the four-day seminars who think they are in for an easy ride may soon have that notion dispelled:

> *"By coming here, you have taken on a*
> *solemn responsibility—*
> *and you can't wriggle out of it."*

Dare I hope that readers of this book will feel a similar responsibility? For:

> *"We've got some big changes to make,*
> *and you're going to have to make them.*
> *Who else will do it?"*

Nottingham, England   Henry R. Neave
14 October 1989

PREFACE TO SECOND PRINTING

Bearing out the earlier statement that "any book on Deming is bound to be out-of-date before it gets published," there have indeed been major further developments in Dr. Deming's thinking since this book was completed, particularly regarding the System of Profound Knowledge. I have therefore taken advantage of the opportunity afforded by the second printing of this book to almost completely rewrite Chapter 18 and make several additional improvements elsewhere. For the benefit of those readers already familiar with the first printing, I have taken care to see that new text is, when possible, of similar length to the text which it replaces, so that the overall structure and pagination of the book are little affected.

19 July 1992   Henry R. Neave

ACKNOWLEDGEMENTS

As we shall emphasise in Chapter 8, any process has a large number of inputs, usually far more than we normally recognise. In Chapter 8, we classify those inputs into six categories. One of those categories is People. Writing this book has, of course, been a process—quite some process! I recognise below some of the People without whom the output from this process would have been all the poorer.

I can begin nowhere else than with my partners in this enterprise, my publishers: SPC Press of Knoxville, Tennessee. I use the term "partners" advisedly (see Chapter 22 on customer-supplier relationships). I have previously authored or co-authored four books, produced by two other publishers, and I have been happy with both those publishers. But, if I was happy with them, I am delirious about SPC Press! They have been truly great. Their enthusiasm, encouragement, guidance, and cooperation have been supreme, coupled with extremely hard work and genuine joy—and fun—in that work. For me also it has been a joy working with them. I must mention by name Frances and Don Wheeler, and Sheila Poling, not forgetting Sharon Farmer and Kristin VanDusseldorp who, in effect, kept the rest of the company going while those first three concentrated their efforts on this book. I am also extremely grateful to John Hart of Hart-Graphics for his care and attention in producing artwork of very high quality.

One aspect of this cooperation with the publishers has been an intriguing mixture of English and American conventions in the book—though heavily weighted toward the English. In fact, the publishers have changed my text only very little: they have still allowed me "programme," "fulfil," and "labour," etc., and all those "esses" instead of "zeds" (sorry: "zees"), though I did relinquish "tyre," "enrol," "aluminium," and "centre," and some punctuation conventions (as you may be able to see in this sentence)—and I even permitted just one "gotten"!

Next, there is of course no way that one can write a book of this nature without borrowing and quoting material from numerous other writers, researchers, speakers, and friends. I have mentioned a number of these in the main text, and there are some detailed attributions in the Permissions section. However (again remembering how numerous are the inputs to the process), any such efforts are bound to be incomplete, so let me now offer universal acknowledgement to everybody who has contributed in any way to my text, explicitly or implicitly, recognised or unrecognised.

There are plenty of other thank-yous that I want to express. The first goes to my cousin Audrey Galloway and her husband Al for their kind hospitality at Fife Lake in Michigan in the summer of 1989, where much of the planning and early work for this book took place. Next, I thank my mother (who is of similar vintage to Dr. Deming!) for her care and concern for me while I was carrying out the most intensive work—including the continuous provision of cups of tea and coffee and repeated reminders for me to go to bed before it was time to get up again! Thanks also to my many friends and colleagues, both inside and outside the British Deming Association (BDA), for putting up with my various absences and latenesses during this busy period.

A number of friends have been kind enough to read an earlier draft of this book, and to provide me with constructive comments and suggestions for improvement. These friends include Kieron Dey, Mike Dickinson, Keith France, Malcolm Gall, Richard Kay, Beryl Moore, Alistair Morrison, Colin Nichols, Eve Williamson, and Peter Worthington. Their help and their contributions have been invaluable. I am also grateful to Mary Tolley for her help at that stage with the diagrams.

More generally, I must thank everybody in the BDA who

has contributed in any way, during the two and a half years of the Association's existence, to my understanding of Deming's work, and for giving me the opportunity to learn more. Special thanks also to my "boss" at the University of Nottingham, Professor Adrian F. M. Smith, for his substantial personal and practical help over the same period.

Something else which happened around two and a half years ago was that Linda Bargent came to work with me, ostensibly as a half-time secretary. She still supposedly works only part-time, but she has become a wholly invaluable colleague and participant and helper in my work. In particular, she performed miracles during September and October 1989, facing the ongoing barrage of dictation tapes with fortitude, and somehow still managing to keep everything else ticking over with her usual efficiency and good humour.

And finally to Dr. Deming. What can I say? I hope he will accept this book as some expression of my gratitude to him. I thank him for reading and commenting in detail on the three papers (mentioned in the Preface) which were the precursor to this book, and for suggesting that I develop them into something more substantial. I thank him for his patient and careful reading of the whole text, and for his numerous suggestions of points of clarification and correction of factual detail. But, far more broadly, I am indebted to him for his continued encouragement, help, friendship, and warmth which he has shown toward me ever since he asked me to assist him at his first London four-day seminar in the summer of 1985. I thank him for adding a new dimension to my life, and likewise to the lives of so many others, both known and unknown to me.

Nottingham, England
3 April 1990

PART 1

PAINTING THE BACKCLOTH

PAINTING THE BACKCLOTH

The book begins with a short chapter, almost entirely in Dr. Deming's own words, on the necessity for the transformation of which he speaks. His outstanding book of 1986 is entitled *Out of the Crisis*. What crisis? Are we in a crisis? Clearly, he considers that we are, and in Chapter 1 he tells us why.

Before we introduce Dr. Deming's philosophy of management, Chapter 2 summarises the history of the man himself. This is not just to tell a good story—though it *is* a good story! The Deming philosophy is *big* and, the more one learns about it, the more massive it becomes. It has been steadily developed over a period of more than 60 years. Some hint, at least, of the history of that development helps us to view the philosophy from a better perspective.

In Chapter 3 we attempt an ambitious task: a summary of many of the rudiments of the Deming philosophy—the 14 Points for Management, the Deadly Diseases of Western-style Management, and the 16 Obstacles to the Transformation. Naturally, this chapter is no replacement for the 130 pages in Chapters 2 and 3 of *Out of the Crisis* where Deming introduces these topics; it is just intended as a stop-gap for those readers who have not yet read Dr. Deming's book. However, it may also prove to be a helpful reference for those already familiar with *Out of the Crisis*.

We conclude Part 1 with an introduction to what may be called *statistical thinking*, which is founded (as is the Deming

3

philosophy itself) on Dr. Walter Shewhart's pioneering work in the 1920s on the nature of variation. An essential feature of this, which we shall see many times in this book, is how *lack* of understanding of variation often leads, despite all the good intentions in the world, to making things worse instead of better. The chapter includes a discussion which highlights the differences between common (and, I believe, inferior) usage of control charts and Shewhart's use of, and need for, this tool which he himself invented.

Chapter 1

DEMING'S VIEW
OF RECENT INDUSTRIAL HISTORY

What are your impressions of the industrial history of this century, particularly of its second half, and your reasons for those impressions?

Here are the views of Dr. W. Edwards Deming. The words in this first chapter are virtually all his, extracted from both seminars and written material. The chapter is written in the style of a personal presentation by Dr. Deming to an audience in America, his home country. However, much of the presentation is also relevant to Britain and other Western countries.

North America has contributed much to new knowledge and to new applications of knowledge. Around the 1920s, the United States contributed to the world interchangeable parts and mass production which put manufactured products into the hands of millions of people the world over that could not otherwise have had them. This country became greater and greater.

Following the devastation of World War II, the United States was the only part of the world that could produce manufactured goods to full capacity. The rest of the industrial world lay in ruins. The rest of the world were our customers, willing buyers. The world needed our products in quantity; they waited in line to leave gold and to take away what we could produce. Gold flowed into Fort Knox. But where is it now? It's not there any more.

Everyone expected the good times to continue and to wax better and better; what could stop them? It is easy to manage a business in an expanding market, and to be hopeful. A lot of people are still living in the 1950s. They are still thinking that because they had breakfast this morning, and have dinner to come, then everything's all right. But, in contrast with expectations, we find on looking back that we have been on an economic decline for decades.

What has happened? It's hard to believe that anything has happened. The change has been gradual, not visible week to week. It is easy to date an earthquake, but not a decline. We can only see the decline by looking back. A cat is unaware that dusk has settled upon the earth. With a cat's eyes, dusk appears as good as midday. But a cat in total darkness is as helpless as any of us. A duck doesn't know it's raining. How would it know?

This country[1] stayed in ascendency until about 1958. But then the world began to change. Around 1958, Japanese goods started to flow in. The price was good and the quality was good, not like the shoddy quality that came from Japan before the war and just after: cheap but worth the price. Preference for at least some imported items gradually climbed and became a threat to North American industry.

Were Americans caught napping? Are we still napping? Markets are now global. People have choice.

Let us think about North America as it is now. How is the United States doing in respect to balance of trade? The

---

[1] i.e. America

6

answer is that we are not doing well; indeed, we are doing very badly (see Figure 1).

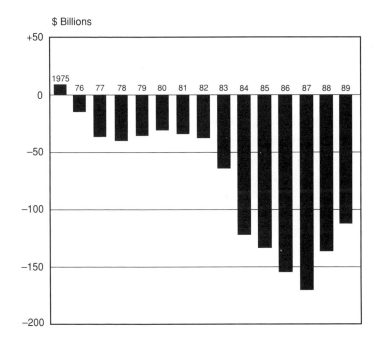

FIGURE 1
U.S. Trade Balance 1975–1989
(U.S. Commerce Department and Census Bureau)

Our deficit in trade, in goods and services, has become an embarrassment.[2] Agricultural products have always brought dollars into the country; wheat, cotton, soybeans, to name a few,

---

[2] Author's note: At the suggestion of my friend Mike Dickinson, until recently of CIBA-GEIGY and now a member of PRISM Consultancy, I also include at the end of this chapter a similar chart (Figure 2) for Great Britain. In the early 1980s we might have felt a little smug comparing our figures with those of America. Now we might not.

still do. There was, until a few years ago, a favourable balance of trade in agricultural products, but no longer. Imports of agricultural products overtook exports in 1986 and, as someone in one of my seminars pointed out, if we could put illicit drugs into the accounting then our deficit in agricultural products would show up worse than the published figures.

The problem is not all to do with agricultural and manufactured products. Some services contribute to our (negative) balance of trade (see *Out of the Crisis*, pages 187–188). For example, the United States pays four times as much for the transportation of people compared with anywhere else in the world.

What can we do? We can try restrictions on trade. What country has the most restrictions on trade? The United States holds this doubtful honour. A few years ago, it was France that held the honour: the United States was only Number 2. Now we are Number 1. Some restrictions erected by the United States are subtle. For example, only 2000 books of a given title, in English, may come in. No one can say that books are excluded from entry, but for practical purposes they are.

Who is our best customer? The answer is that the province of Ontario is our best customer. I read something recently on the biggest export-dollar earners. General Motors was Number 1, Ford was Number 2, the aircraft industry was Number 3. It didn't make sense. On looking further, I found that exports include subassemblies, parts for assembly, many of them going to Ontario. Later, they come back to this country. But they count as exports.

Our problem is quality. Can't we make quality? Of course we can. And some American products are superior. We are thankful for them. Unfortunately, some good American products have little appeal beyond our borders: good paper clips, for example.

Some industries are doing better than ever. There are more automobiles in the United States than ever before, and more people travel by air. Do such figures mean decline or advance?

An answer would have to take into account that in 1958 we had inter-city trains. There was choice: air, train, or automobile. Now we have only limited train service. So we have to go by air or by automobile.

One of our best exports, one that brings in dollars, is materials for war. We could greatly expand this income but for moral concerns. Another big earner of dollars is scrap metal. We can't use it, so we sell it. The raw material in this microphone is worth maybe 15 cents. We sell it for that, and buy it back at a seller's price of perhaps $2000. Close on to scrap metal is scrap cardboard and paper. Timber brings in dollars. Timber is important, renewable. American aircraft have about 70% of the world market, and bring in huge amounts of dollars. Equipment for construction is an important export, so I understand.

American movies, a service, bring in dollars. Banking and insurance and other services are important, but nowhere near as much as previously. The British have always been better at them. The eight biggest banks in the world now are Japanese, the ninth is French, and the tenth is American. The next American bank is Number 36. Banking is now important in America mostly for losses on bad loans.

23% of real estate in Washington is owned by the Japanese. The figure in Los Angeles is 44%. There was one time when I awoke in Los Angeles, looked out of the window, and thought I was in Tokyo. A delegate at my seminar in London spoke of "selling off the family silver."

We have been blessed with natural resources. But most of them are nonrenewable. When they have gone, what will we eat? We ship out, for dollars, iron ore (partially-refined), aluminum, nickel, copper, coal: all nonrenewable stock. Scrap metal is nonrenewable.

Have we been living on fat? We have been wasting our natural resources and, as we shall see, destroying our people. We need them.

What is the definition of a colony? A colony exports raw materials and buys costly manufactured products and service. Have we become a colony?

> "What is the world's most underdeveloped nation? With the storehouse of skills and knowledge contained in its millions of unemployed, and with the even more appalling underuse, misuse, and abuse of skills and knowledge in the army of employed people in all ranks in all industries, the United States may be today the most underdeveloped nation in the world."

(*Out of the Crisis*, page 6).

What are we to do? I believe we must start out afresh with confession: not of sin, but of stupidity. What we have been doing, hour by hour, is wrong. We have been doing everything possible to eliminate profit, and have for three decades. Does anybody really care about profit? I see very little evidence of it. If so, surely he[3] would learn more about it. The chance of successfully improving our position in the future under the present system of management is nonexistent.

---

[3] Author's note: Like most writers in English, I greatly regret the fact that there is no convenient method in the language of referring to both male and female in the singular third person. So I must ask the reader to please interpret, wherever appropriate, "he" as "he or she," "him" as "him or her," etc., rather than expecting me to cover the text with such inelegant constructs.

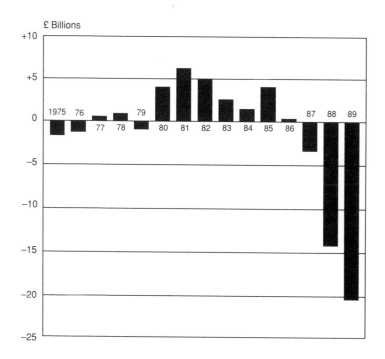

FIGURE 2
British Trade Balance 1975–1989
(Based on data mostly from *Monthly Digest of Statistics*)

Chapter 2

A BRIEF BIOGRAPHY

William Edwards Deming was born on 14 October 1900. Not surprisingly therefore, his story is a long one! But it is important to learn something of that history in order to gain a better perspective on Deming's work. The unwary may feel deluged by a confusing plethora of advice and requirements in Deming's teachings; and who, to be fair, could be surprised if some newcomers regard only part of those teachings as reasonable, but others as rather peculiar, and some as downright fanciful?

However, to barge into Deming's work unprepared is like looking at tips of icebergs. We need to have a strong sense that what we hear and read now is the product of well over 60 years of active thought, development, and practice on the part of this brilliant and far-sighted individual, aided by the creativity of countless others in industry, education, statistics, physics, psychology, etc., whose work he has diligently studied, sifted, and selectively adopted. And, rather as he counsels us to focus attention not on the (downstream) output of processes but on the (upstream) causes of that output, so we need to study the background and fundamental building blocks that have brought the Deming philosophy to what it is today. In addition, while Deming himself warns against having our thinking too much influenced by "results," it can be helpful to many people to have some idea of

13

the extraordinary track-record of this nonagenarian "consultant in statistical studies."

Let us not go back to the very beginning just yet, but pick up the story some 15 years ago. At that time, the name of Dr. W. Edwards Deming was not widely known—at least, not on this side of the world. Of course, it was not wholly unknown either; but most of those who had heard of Deming recognised him more as a distinguished academic and consultant statistician[1] than as somebody with important things to say about management. Even in those parts of the statistical community in which he was well-known, it was not (as he recently wrote to me) "for excursions into new worlds." And, if the statistical community was slow to recognise the far-reaching importance and potential of Deming's work, there is no doubt that the industrial community was even slower: the industrial community in the Western world, that is.

One American industrialist who heard of Deming's name in 1974, and was somewhat mystified by the circumstances in which he heard it, was William E. Conway, then President of the Nashua Corporation and later to become its CEO (Chief Executive Officer). The Nashua Corporation is a Fortune 500 company, situated in the town of Nashua, New Hampshire, whose main lines of business include hard disks for computers, copy machines, photo-processing, and various paper products.

In the early 1970s, Nashua was involved in a consortium of some five American companies, along with a German chemical company and the Japanese manufacturing company, Ricoh, developing a new, potentially world-beating copy system. Work had started in 1969, and the system was scheduled to go onto the market in 1974. Because the product was to be manufactured in Japan, most of the big technical meetings were held in Tokyo.

On returning from one of these visits, Nashua's Director

---

[1] A comprehensive summary of Dr. Deming's numerous publications is included in Chapter 15 of Cecelia Kilian's biography: *The World of W. Edwards Deming.*

of Research and Development had an unusual story to relate. Instead of being received with normal Japanese courtesy, the American visitors had been virtually ignored. A couple of meetings did eventually take place, but only at extremely unusual and unsocial hours. Naturally, when he heard of this, Conway wanted to know what was going on. Here, in Conway's own words, is what his R&D Director told him:

> "They're all nuts over there. They're all collecting data on stuff, plotting them on all these little charts, and then they're all going round fixing everything."

Conway asked: "Who's doing all this?" The answer was:

> "Everyone's doing it: the President, the Vice-President of Sales, the Controller, the R&D chemists, the hourly workers, the foremen, the accountants. Everybody is doing it, Bill; they're all doing it!"

And why were they "doing it"? In 1969, Ricoh had embarked upon a five-year programme in pursuit of the prestigious industrial award known as the Deming Prize. The Nashua team was unfortunate enough to have gone to Tokyo just before the deadline for Ricoh's submission for the Deming Prize.

By the late 1970s, Nashua was competing against some of the leading Japanese export companies in many products and services. Conway saw the substantial differences in prices which the Japanese competition was able to charge. Of course, there was nothing new in Japanese prices undercutting those of Western companies. What was new was that, instead of the "cheap and shoddy" image many of us remember from our formative years, not only were the prices low but the quality, consistency, and reliability were now superior. Why? How could this be?

Of course, Conway had heard of the Deming Prize back in 1974, and he began to notice more and more mention of the name on his repeated visits to Japan. He became aware of the high respect and, indeed, reverence that the Japanese had for this Dr. Deming. He discovered that the Deming Prize ceremonies were shown on prime-time television, and found that the companies which had won the Deming Prize proudly displayed huge brass replicas of it in their main entrance halls.

Even so, Conway did not act upon this awareness for nearly five years. Then, in early 1979, one of Conway's colleagues at Nashua recalled that, during the 1950s, his previous employer had engaged a Dr. Deming as a statistical consultant. Conway told him to find out if Deming was still alive and, if so, to try to locate him. On Tuesday 6 March 1979, Conway telephoned Dr. Deming at his Washington home and asked him to come and visit. Deming came on Friday of that same week.

Deming, then 78, had an incredible story to tell. He told Conway of some of the things he had done in America, particularly just before and during World War II. He told of how, in the post-war boom years, most American ears became very deaf to his message. He told of how he eventually found himself talking to some Japanese executives in mid-1950. And he told of what had been happening in Japan since then.

Conway, much against the advice of his senior colleagues, asked Deming to consult for Nashua. And Deming refused! At least, he said that he was not interested unless Conway made a very special commitment. The commitment was that Conway himself, as Chief Executive Officer, must become the leader and agent of the change that Deming required, for that responsibility could not be delegated. In earlier years, Deming had had quite enough disappointments from American managers, and he was determined not to make the same mistakes again. Conway agreed. Deming's Western renaissance had begun.

As Conway proceeded to learn and understand Deming's teachings and get them accepted in his company, and as (in spite of Deming's warnings against expecting too much in the short term) he soon saw exciting and valuable things beginning to happen, he became convinced of the need for Deming's work to become much more widely known and understood nationally.

Conway himself did much to spread such awareness, speaking to his fellow-executives around the country at every opportunity. One such opportunity arose on 11 October 1980. The occasion was a panel discussion on productivity at the National Association of Manufacturers' Board of Directors meeting held in Phoenix, Arizona. Some excerpts from Conway's speech typify his enthusiasm and belief in what Deming had done and could do:

> "In my company, thanks to the teaching of Dr. W. Edwards Deming, we now realise that increases in productivity can be achieved daily through greater emphasis on quality in all phases of our operations. Dr. Deming has worked with Japanese industry for 30 years. He is credited with major contributions to its giant strides in quality and productivity. There is a direct correlation between the two: as quality goes up, so does productivity. Dr. Deming's approach is based on the statistical control of quality to bring about a new way of managing your business. Dr. Deming is the Father of the Third Wave of the Industrial Revolution. The Japanese manufacturers utilising the statistical control of quality are sweeping the world in the second half of the 20th Century, just as American manufacturers utilising mass production swept the world in the first half."

It is worth mentioning that, prior to meeting Deming, Conway knew next to nothing on the subject of statistics. He confesses that he vaguely remembered something about a "bell-shaped curve" in one of his college courses, but that was about the size of it. He continued:

> "For years the Japanese have been using statistics and charts to measure, evaluate, and improve

their operations. They use statistical information—facts, figures, charts—to help to make the continual improvements that they know will lead directly to higher productivity and lower cost with consistent quality suited to the marketplace. This is the kind of competition faced by many of us here today. It represents a formidable level of quality and productivity. And don't for a moment dismiss the Japanese success as one that grew out of Japan's unique cultural heritage or low wages and therefore would not apply to American manufacturers. *The Japanese success is based upon the statistical control of quality, introduced to them by an American.*"

It would be wrong of me to give the impression that Conway did everything right. I believe he failed to grasp some of the deeper implications of what Deming was saying to him, and he did make mistakes. But who wouldn't have done, in his position? The important thing is that he did recognise something of the momentous nature of Deming's work, and he strove to get the man, and what he understood of his message, much better-known. For that, we owe Bill Conway a considerable debt of gratitude.

Whereas Conway worked hard at getting senior managers to learn about Deming, initial awareness over a much broader spectrum was created on 24 June 1980, as the result of an NBC television documentary: *If Japan Can, Why Can't We?* Clare Crawford-Mason, producer of this documentary, had had a hard time finding suitable material. That was until somebody pointed her toward Deming. Deming, she was told, "had done a lot of work in Japan." What happened then is well-related in Mary Walton's book, *The Deming Management Method*:

"Crawford-Mason contacted Dr. Deming, who invited her out to talk. He spoke of his work in Japan, and showed her yellowed clippings of stories the Japanese had written. Crawford-

Mason didn't know what to think. He was nice, if
eccentric; he reminded her of her father; but
what he said, if true, was astonishing. 'He kept
going on and on and on that nobody would listen
to him.' Their first conversation led to five inter-
views, consuming 25 hours. The more they talked,
the more impressed she was, and the more sus-
picious she became. It was simply incredible.
'Here is a man who has the answer, and he's five
miles from the White House, and nobody will
speak to him.' She contacted a high-ranking economics official from the Carter Administration
and asked if he knew Dr. W. Edwards Deming.
He didn't."

By this time, Deming was able to tell Crawford-Mason
about his recent involvement with Nashua. She visited the company, and it became the location for the final section of her documentary, featuring both Deming and Conway. Many of the people whose careers are now centered on Deming's work first heard
of him through that telecast. Mary Walton continues:

"The next day (i.e. 25 June 1980), the telephone
rang relentlessly in Dr. Deming's basement office.
'We were bombarded with calls,' recalled his
secretary, Cecelia Kilian. 'It was a nightmare.'
Many of the callers sounded desperate. 'They
had to see him tomorrow, or yesterday, or their
whole company would collapse!' "

Now that we have reached relatively recent history, it
is both impossible and unnecessary to catalogue all that has happened since—there is so much. In any case, the reader is likely to
be more familiar with recent rather than earlier happenings.
But here, for completeness, is a brief sketch of some of the events
which have happened since Conway first met Deming in March
1979.

Later that same year, with Conway's encouragement, Deming started giving his now-celebrated four-day management seminars. Initially, audiences sometimes numbered only a dozen or two. After a couple of years, audiences were measured in hundreds rather than dozens. He still gives such public seminars about 15 times a year, frequently with capacity audiences. Mrs. Kilian told me of one seminar in 1985 at which the audience exceeded 1000; however, he has preferred to keep attendances down to around 500 since then.

It is clear then that Deming still reaches several thousands of American managers each year through his seminars, and it would be difficult indeed to estimate the number that have shared this experience through the decade, especially if private in-company seminars are also included. Further, unlike in the early 1980s, Deming now attracts many senior managers; I recall one occasion in the summer of 1987 when the 52 top managers of just one company were in attendance.

Of the major corporations in America, Ford in particular has "gone public" with its commitment to the Deming philosophy. Don Petersen, who recently retired as Chairman of the Board of Ford Motor Company, was on record long ago as saying that he was "proud to be called a Deming disciple." Mary Walton's two books include case histories from, besides Ford, AT&T, Bridgestone Inc., Campbell Soups, Florida Power & Light,[2] Globe Metallurgical, Honeywell, the Hospital Corporation of America, Janbridge, Inc., Malden Mills, Microcircuit Engineering Corporation, and the US Navy. Other companies which soon became interested in Deming's approach were Kimberly-Clark, Procter & Gamble, and Velcro. Other public utilities have become involved, and some famous work has been carried out within the city government of Madison, the Tri-Cities area of Tennessee, and

---

[2] Florida Power and Light has become the first non-Japanese recipient of Japan's Deming Prize (in 1989).

the state government of Wisconsin.

Numerous Deming Users Groups—organisations dedicated to spreading awareness and understanding of Deming's work, and to helping interested organisations put his teachings into practice—have sprung up all over America. In the early 1980s, the MANS[3] Foundation in Holland and the W. Edwards Deming Institute of New Zealand were formed for similar purposes. In November 1987, with the help of many fine friends, I launched the British Deming Association (BDA), and we were delighted when Jean-Marie Gogue founded the Association Française Edwards Deming early in 1989.

During the 1980s, Deming has received numerous awards, honorary doctorates, and medals, including the National Medal of Technology from President Reagan in 1987. The citation for that Medal honours Deming for his "forceful promotion of statistical methodology, for his contributions to sampling theory, and for his advocacy to corporations and nations of a general management philosophy that has resulted in improved product quality with consequent betterment of products available to users, as well as more efficient corporate performance."

In somewhat startling comparison is the fact that, as long ago as 1960, Deming was decorated in the name of the Japanese Emperor with the Second Order of the Sacred Treasure. He was the first-ever American to receive such an honour.

Having gotten up-to-date on recent history, let us now sketch the first several decades of Deming's life. The interested reader will find a much more comprehensive and absorbing account in Mrs. Kilian's biography (already referenced in a footnote near the beginning of this chapter).

William Edwards Deming's earliest years were spent in the state of Iowa, first in Sioux City, and then on a farm near Polk

---

[3] Management en Arbeid Nieuwe Stijl (Management and Labour New Style); now the MANS *Association*.

City owned by his maternal grandfather. The family moved to Wyoming when he was seven, and he obtained his first degree, in Electrical Engineering, at the University of Wyoming in 1921. He remained at Wyoming one further year for additional studies in Mathematics, also teaching Engineering classes.

He taught Physics the following two years at the Colorado School of Mines, and enrolled at the University of Colorado for a Master's Degree in Mathematics and Physics. He entered Yale in 1925, and was awarded his Ph.D. in Mathematical Physics in 1928. A year earlier he had joined the United States Department of Agriculture as a mathematical physicist, and he stayed with USDA until early 1939.

Statisticians will be aware that much of the work underlying their subject as they know it today was being developed in the 1930s. In particular, considerable progress was made around that time in experimental design, with fruitful applications to agricultural problems. So it is not surprising that Deming's interest in probability and statistics flourished; and in 1936 he came to London to study under the "Father of Statistics," R. A. (later Sir Ronald) Fisher, at University College.

But, in spite of his keen interest in the mainstream developments of Mathematical Statistics at this time, Deming had found even greater inspiration in the work of Dr. Walter Shewhart, the originator of the concepts of controlled and uncontrolled variability, statistical control of processes, and the related technical tool of the control chart. Deming first learned of Shewhart's work through taking summer vacations jobs in 1925 and 1926 at Western Electric's Hawthorne plant in Chicago. He did not meet Shewhart until the autumn of 1927, but the two spent much time working together thereafter.

Shewhart's fundamental work was published in a number of papers and then in his 1931 book: *Economic Control of Quality of Manufactured Product*. In 1938 Deming persuaded Shewhart to give a series of four lectures at the Graduate School of the Department of Agriculture, and the content of those lectures was published in 1939 under the title: *Statistical Method from the Viewpoint of Quality Control*. Deming's supreme respect for Shewhart's work is clearly seen in the following extract from Mrs. Kilian's biography:

> "A half century has passed by since Dr. Shewhart's great book of 1931 appeared, and almost half a century since his book of 1939 appeared. Another half century will pass before people in industry and in science begin to appreciate the contents of these great works.
>
> One can say that the content of my seminars ... and the content of my books ... are based in large part on my understanding of Dr. Shewhart's teaching. Even if only ten per cent of the listeners absorb part of Dr. Shewhart's teachings, the number may in time bring about change in the style of Western management."

At the root of Shewhart's work was the realisation that there are, in general, two kinds of variability in the output from a process. There is *controlled* variability which is due to the process itself: the way it has been designed, built, set up, the way people have been trained to work in it, and so on. And there is *uncontrolled* variability from sources outside the process, which prevent it from performing as well as it could if it were so permitted. Deming refers to the two types of variability as being due to "common" causes and "special" or "local" causes respectively. An introduction to Shewhart's fundamental work, and its implications, is given in Chapter 4.

Whereas Shewhart concentrated mostly on production processes, Deming realised that his ideas were just as applicable in nonmanufacturing environments, e.g. in administration, service operations, budgeting, forecasting, etc.. Indeed, some of the most fruitful applications that I have seen have been in administration departments.

In 1939, Dr. Deming began work at the National Bureau of the Census, and almost immediately started applying the concepts to routine clerical operations, such as coding and cardpunching, in preparation for their use in the 1940 Population Census. Through getting such processes into statistical control, and then accordingly being able to improve these processes, there was a huge decrease in the need for inspection, verification, etc.. As a result, it is on record that productivity in some of these processes increased by as much as *six times*. Savings were of the order of several hundreds of thousands of dollars (an incredible amr int, realising the value of money so long ago), and the Census results were published much earlier than usual.

In 1942 Deming, at the instigation of fellow-statistician W. Allen Wallis (at that time on the staff of the Statistics Faculty of Stanford University and later to become Under-Secretary of State for Economic Affairs) set up courses to teach his and Shewhart's methods to industrialists, engineers, designers, etc., who were particularly involved in the war effort. The programme had a strongly beneficial effect on both the quality and the volume of production; there were spectacular reductions in scrap and the need for rework. Some of the people involved in the teaching of this programme were primarily involved in setting up the American Society for Quality Control in 1946.

However, the advances made during the war were not subsequently sustained. Producers found themselves in a boom sellers' market: essentially anything that they produced was readily saleable. Why then bother with quality and statistics and all that nonsense? Actually, Deming does not believe that he ever really got his ideas through to management, even during the war. Quoting him from another biography, Nancy Mann's *The Keys to Excellence*:

> "The courses were well-received by engineers, but management paid no attention to them. Management did not understand that they had to get be-

hind improvement of quality and carry out their obligations from the top down. ... Any instabilities can help to point out specific times or locations of local problems. Once these local problems are removed, there is a process that will continue until someone changes it. ... Changing the process is management's responsibility. And we failed to teach them that."

After the war, Dr. Deming was invited on two occasions by General MacArthur to go to Japan as an advisor to the Japanese Census. Whilst there, he met some of the members of JUSE (the Union of Japanese Scientists and Engineers), which had been formed in 1946 by Mr. Kenichi Koyanagi with the task of helping the industrial reconstruction of Japan after the devastation of the war. The Japanese were also later visited by a delegation from Bell Telephone Laboratories, who showed them some of the quality control techniques the Americans had been using during the war.

In a paper published in *Quality Progress* in 1985, Kenneth Hopper tells how Japanese executives in communications companies were also introduced to quality control ideas in 1948 and 1949 through a course presented by Charles Protzman and Homer Sarasohn. When Deming's involvement in the development and teaching of quality control became apparent to some members of JUSE, Kenichi Koyanagi wrote to invite him back to Japan to lecture to them on statistical methods for industry. Deming regards it as highly important that the initiative came from the Japanese, rather than through him trying to impose his ideas on them:

> *"I rang no doorbells;*
> *I did not ask to go."*

Deming went in June 1950. Some 225 people attended his first lecture in Tokyo, and there were 85 in Hakata, 110 in Fukuoka, and 150 in Osaka.

Deming was, of course, very conscious by now of the failure of his ideas to take root in America. As indicated above, he believed that he now knew the reason, and realised he must not make the same mistake in Japan. Again from Nancy Mann's biography:

> "They were wonderful students, but on the first day of the lectures a horrible thought came to me: 'Nothing will happen in Japan; my efforts will come to naught unless I talk to top management.' By that time I had some idea of what top management must do. There are many tasks that only the top people can perform: consumer research, for example, and work with vendors. I knew that I must reach top management. Otherwise it would just be another flop as it had been in the States."

Accordingly, Deming asked his hosts if it would be possible for him to speak to some senior managers. Ichiro Ishikawa, President of JUSE, arranged for him to meet with members of the Kei-dan-ren, the association of Japan's chief executives. The first such meeting was held on 13 July with the presidents of 21 of Japan's leading companies. In all, Deming spoke to some 100 Japanese senior managers during that summer, and in 1951 he reached about 400 more. In December 1950 the JUSE Board of Directors formally resolved to create the Deming Prize "in commemoration of Dr. Deming's contributions to Japanese industry and for encouragement of quality control development in Japan."[4]

The Japanese listened, astonished, as Deming told them that if they adopted his methods then they would begin to capture world markets in a very few years. But Deming was no longer talking just statistics. He was now talking in terms related to what we today might refer to as "Total Quality" or "Company-

---

[4] This quotation comes from a JUSE booklet: *The Deming Prize*, which is reproduced as Chapter 10 of Mrs. Kilian's book.

Wide Quality." Deming himself does not use these expressions; he describes his work simply as "Management for Quality." He was telling them that:

> ## *"The consumer is the most important part of the production line."*

and he was recommending that they start work with their vendors to obtain reliable and good quality supplies. In short, he was telling them so much of what most people in the Western world have begun to think about only in very recent years.

Of course, most of the Japanese did not really believe that Deming's optimism was justified. But they did realise that his ideas made a lot of sense, and that they constituted a very different approach from that which they were currently following. They had great respect for this foreign scholar, and also liked the way that he (unlike some of his fellow-countrymen) treated them with respect and humanity. And, considering their current situation, what did they have to lose? So they embraced the Deming philosophy, and did so with an enthusiasm and dedication almost beyond Dr. Deming's belief:

> ## *"They changed the economy of the world."*

The rest, as they say, is history.

Chapter 3

THE 14 POINTS, THE DISEASES, AND THE OBSTACLES

Dr. Deming's management philosophy is based on an all-embracing concept of quality and an understanding of variation (closely linked to the statistical control of processes), combined with a vital third component dedicated to establishing an environment in which those two seeds may grow and flourish. There is no doubt that Deming's thinking on this third component has been partly influenced by his work with the Japanese. A brief way of expressing his thinking is "management by positive cooperation" as opposed to "management by conflict." He has recently been concisely referring to this as "Cooperation: Win-Win" in preference to "Competition: Win-Lose."

It is interesting to note that Lloyd Dobyns, the narrator in *If Japan Can, Why Can't We?* (see Chapter 2), made the point near the end of the programme that Deming had had the advantage in Japan of a more cooperative management and working environment. Dobyns seemed to have the impression that, if the Deming approach were to succeed in the West, it would have to be adjusted to fit into our confrontational scene. But that could never be. Deming talks of "total transformation of Western style of management"—and he means it! A substantial part of his philosophy is thus concerned with the transformation from an environment of internal competition to one of total and genuine teamwork within an organisation. This is why he says, for example, that:

> ### *"Best efforts are not enough; best efforts will not ensure quality."*

Everybody putting forth best efforts from his own individual viewpoint results in much wasted labour; everyone needs to pull in the same direction, the direction that is of greatest benefit to the company as a whole. If the energy wasted in confrontation between management and workers can be channelled into cooperative effort, it does not take much imagination to visualise how enormous the difference can be. And be clear that this is not the same thing as "tough" management—nor government of left or right—beating (metaphorically, at least) workers into submission. Unlike some other "quality gurus," Deming has in no way been content to try and make do with traditional management *milieu*; he knows that that in itself constitutes an impenetrable barrier to the kinds of improvements which would otherwise be possible. And, of course, when we suffer competition which is not obstructed by such a barrier, the consequences are likely to be all too apparent.

Naturally, there is in the Deming philosophy, as in all other modern approaches to quality, concentration on the customer. We have already seen how, back in 1950, Deming taught the Japanese that they needed to regard the consumer as "the most important part of the production line." But Deming goes much further than other approaches, which tend to speak in terms of "satisfying the customer at the lowest possible cost." For example, we read in *BS5750: A Positive Contribution to Better Business*[1] that:

---

[1] The British Standard "Quality System" BS5750 was the forerunner of, and is mostly identical to, the International Standard ISO9000.

> "Quality has a number of different meanings,
> but BS5750 looks at it in the fitness for purpose
> sense; is the product designed and constructed to
> satisfy the customer's needs?"

This hardly touches on Deming's thinking. On page 141
of *Out of the Crisis* he writes:

> "It will not suffice to have customers that are
> merely satisfied. An unhappy customer will
> switch. Unfortunately, a satisfied customer
> may also switch, on the theory that he could
> not lose much, and might gain. Profit in
> business comes from repeat customers, customers
> that boast about your product and service, and
> that bring friends with them."

Deming also speaks frequently on the need for *staying
ahead* of the customer. The customer does not know what he will
need in one, three, five years from now. If you, as just one of his
potential suppliers, wait until then to find out, you will hardly
be ready to serve him.

The purpose of this chapter is to provide a compact introduction to Deming's 14 Points for Management and to the Deadly
Diseases and the Obstacles to Transformation. It is intended particularly for those who have not yet read *Out of the Crisis*, and so
is essentially concerned with the thinking around the time that
Dr. Deming's book was written. There will be no attempt here to
bring the reader up-to-date on Deming's more recent thinking, in
particular concerning the 14 Points; that will be the task of the
final Part of this book.

However, I am beginning with something written slightly after *Out of the Crisis*: a superb article by my good friends
Peter Scholtes and Heero Hacquebord[2] from Joiner Associates Inc.
of Madison, Wisconsin. Their article, entitled "A Practical

---

[2] Heero Hacquebord is now an independent consultant.

Approach to Quality," commences with a number of "Quality Guidelines." The very first of these brilliantly reflects and develops Deming's appreciation of the prime role of the customer. It is well worth reproducing this, and the subsequent commentary, in full:

"Quality Guideline 1:
Quality Begins with Delighting the Customer

Customers must get what they want, when they want it, and how they want it. An organisation must strive not only to satisfy the customers' expectations. This is the least one should do. A company should also strive to delight their customers, giving them even more than they imagined possible. Your bosses may be ecstatic, the Board of Directors blissful, and your company may be considered a legend on Wall Street. But if your customers are not delighted, you have not begun to achieve quality."

Another contribution from Joiner Associates, which I find very helpful, is Brian Joiner's own "Joiner Triangle"—see Figure 3. As an eight-word bedrock of the Deming philosophy, especially as it was understood at the time *Out of the Crisis* was written, I think it is hard to beat.

The top vertex of the Triangle is "Obsession with Quality." "Obsession" is a striking choice of word. It certainly conveys the profound and primary importance of quality, as opposed to its familiar role of playing second fiddle to more pressing short-term considerations and crises. However, "obsession" also gives the impression of a preoccupation beyond rhyme or reason, and this may be misleading. There *is* rhyme and reason for having an obsession with quality, and it is comprehensively expressed in Deming's "chain reaction" which we learn (as early as page 3 in *Out of the Crisis*) "was on the blackboard of every meeting with top management in Japan from July 1950 onwards."

This chain reaction highlights the fact that cost reductions, business success, and increased profitability are natural consequences of improvement in quality in the sense with which we understand and develop the meaning of "quality" in this book.

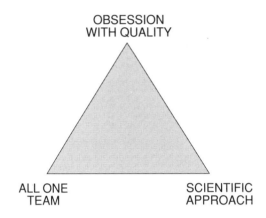

FIGURE 3
The Joiner Triangle
(Joiner Associates)

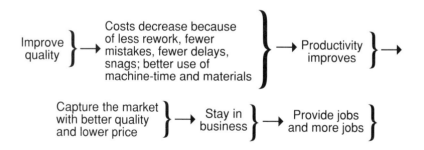

FIGURE 4
Deming's "Chain Reaction"
(Dr. Deming, *Out of the Crisis*, page 3)

33

The Joiner Triangle shows that such all-embracing quality is achieved by the coalescence of two forces: total teamwork and the "Scientific Approach." The Scientific Approach requires deep understanding of the nature of variation, particularly its division into controlled and uncontrolled components due respectively to common and special causes. This is a very crucial aspect of what Deming now refers to as "Profound Knowledge" (see Chapter 18), and it affects all facets of management. For it is only by correctly diagnosing the most important sources of variation, and then reducing or even eliminating them, that quality (reliability, consistency, predictability, dependability) can be improved.

The Scientific Approach calls for decision-making and policy-making on the basis of good information, both numerical and non-numerical—not just by "gut-feel" or mere short-term considerations. It will often include the analysis of data by statistical techniques and SPC (Statistical Process Control)—but it also includes knowledge and understanding of the *limitations* of such techniques, and awareness of the crucial importance of phenomena which cannot be quantified (see Chapter 10). Indeed, Deming's fifth "Deadly Disease" of Western management is headed: "Running a company on visible figures alone (counting the money)" (*Out of the Crisis*, page 121). Deming frequently quotes Dr. Lloyd Nelson,[3] the Director of Statistical Methods in the Nashua Corporation:

> "The most important figures needed for management of any organisation are unknown and unknowable."

(*Out of the Crisis*, page 20).

---

[3] It was through Lloyd Nelson's invitation for me to help the British subsidiaries of the Nashua Corporation in 1981 that I first learned of Dr. Deming's work; and for that, of course, I shall be forever indebted to him.

Those who find such statements surprising (especially from statisticians!) have not really begun to understand Deming's teachings. But surely Lloyd is correct. Who can know the "figures" reflecting the results of delighting your customer rather than merely satisfying him? Not only does he come back for more—without need of advertising or other persuasion—but he is likely to recommend you to his colleagues and friends, both inside and outside his company. What figures could you use to represent Japan's obsession with quality over recent decades? What figures can you use to show the increased value of a happy and intrinsically-motivated worker as opposed to one who is only there for his take-home pay (see Part D, Items 5 and 6 in the System of Profound Knowledge in Chapter 18)? And what figures can quantify the harm caused by letting down your customer in any way or by having a disillusioned and demoralised work-force?

The contribution of the Scientific Approach to the Obsession with Quality therefore must go way beyond dealing with (knowable) figures. Topics which can be included under this general heading take up a considerable proportion of this book, in particular the next chapter, all of Part 2, and Chapter 16.

The remaining vertex of the Joiner Triangle is "All One Team." Teamwork, genuine full-hearted teamwork, when it exists, contributes so much in work—as it does to life in general: in the home, in committees, on the sports-field, in music-making. I do not believe that the success of a football team is ruled by the sum[4] of individual abilities of its members. Of course, that sum is an important part of the formula, but a group of talented individuals is often beaten by a *team* of players for whom that sum is less. In years gone by (when I had time for hobbies!) I used to be actively involved in local musical organisations, particularly as Musical Director of amateur operatic societies. That was an excellent way to learn the need for teamwork—on stage, behind the scenes, and in the orchestra pit. Dr. Deming has also enjoyed a

---

[4] i.e. calculated just by adding together

lifelong love of music, so it is no surprise that he draws the same parallel with regard to orchestras:

> *"The players are not there to play solos*
> *as prima donnas,*
> *each one trying to catch the ear of the listener.*
> *They are there to support each other.*
> *They need not be*
> *the best players in the country."*

Indeed, there have been occasions when, for special celebratory or commemorative events, leading soloists have congregated to play together. The result is always interesting, but it is not necessarily easy on the ear.

In work, "All One Team" is rendered impossible by many modern management practices, such as Management by Objective (MBO), annual performance appraisals, and the use of arbitrary numerical goals and targets—all of which foster competition and conflict between people, and sometimes between whole departments, as opposed to their working together for true benefit of the company. Several well-intentioned quality-improvement notions from other sources constitute obstacles which are just as serious: examples are Cost of Quality and Zero Defects. Even concepts which are essentially good, such as Quality Circles and Just-in-Time, can do more harm than good if used in an unsuitable environment.

Again, nobody can approach true comprehension of Deming's message without understanding what he is saying on these matters. Such management practices have become popular (at least amongst management!) because they *make the best of a bad job*—they make things less bad than they would be otherwise. What is unseen by those who do not know or understand Deming's teachings is that these practices also constitute massive obstacles to the needed transformation.

Deming's 14 Points have developed gradually during a period of over 20 years. When he began putting them in writing, there were rather less of them—for that was when his principal audience was still the Japanese and they did not have to be told to "Drive Out Fear" (Point 8) or to "Permit Pride of Workmanship" (Point 12). The number reached 14 when Deming started to present his four-day seminars in America some ten years ago.

The 14 Points (sometimes he calls them "Principles" or "Obligations" of management) are not written on tablets of stone: indeed, there have been many minor, and a few quite major, adjustments to them during the decade, reflecting the way that Deming sees the world changing and the changing needs of the people with whom he works. Indeed, early in 1990 he issued for discussion a whole new set of enhancements and refinements to the 14 Points. Although my colleagues and I found these to be very helpful, he did not make them "official"; consequently, to avoid confusion, they are not explicitly included in this volume. Their existence is mentioned here to demonstrate his continued flexibility and responsive attitude to the changing industrial world.

The version that is used here is derived from at least six different versions that I had seen in or before 1986. The words are virtually all Deming's, and they are presented here in a compact format which I hope will form a helpful introduction for newcomers to Deming's work. The small-print comments following each Point are mine.[5]

A few words of warning before we embark upon the 14 Points. The Points do not constitute the whole of the Deming philosophy, though they are especially important constituents of it. They are not a list of instructions; they are not techniques; they are not a check-list. They are vehicles for opening up the

---

[5] I am grateful to Colin Nichols, until recently of Austin-Rover and now a director of PMI, for suggesting a number of additions to these comments.

mind to new thinking, to the possibility that there are radically different and better ways of organising our businesses and working with people. Certainly, any full adoption of the Deming philosophy will require constant attention and movement toward the principles expressed in the 14 Points.

But there is great danger in simply obeying the words without first studying and developing a deep understanding of why he is saying these things. Indeed, I would suggest that to treat the 14 Points just as a recipe is, in fact, a pretty sure recipe for disaster! I would not recommend anyone to try to adopt any of the 14 Points before they have gained that deep understanding. Only then can they have the ability to judge how the transition, the "total transformation," should be attempted in the particular circumstances of their own organisation. For the need is not simply to adopt the 14 Points, individually or collectively, but to create a new environment which is fully consistent with and conducive to them. And this is not a project, nor a programme. This is never-ending—forever.

The reader may care to keep the Joiner Triangle in mind as he works through the 14 Points. As one gets familiar with the concepts, it becomes more and more apparent that all three vertices of the Triangle have a bearing on every one of the 14 Points. However, perhaps the most obvious connections are as follows:

Obsession with Quality:	Points 1–6, 13–14
All One Team:	Points 7–9
Scientific Approach:	Points 10–12

I hope this observation will help the reader to appreciate that the 14 Points, which may at first glance appear rather diverse and autonomous, are in fact all interlinked within a strongly-coordinated philosophical structure.

THE FOURTEEN POINTS

1. CONSTANCY OF PURPOSE

Create constancy of purpose for continual improvement of products and service, allocating resources to provide for long-range needs rather than only short-term profitability, with a plan to become competitive, to stay in business, and to provide jobs.

It is fruitless to accept Deming's philosophy in principle, but then forget it in practice. What sometimes happens is that management indicate their agreement, but then allow virtually anything else (i.e. all the "old problems") to take priority. There must be a consistent, inexorable, never-ending, widespread push for continual improvement in all activities and operations in the company. People are so used, these days, to seeing new management gimmicks every few weeks—which often disappear as quickly as they came. It will need time, with such a history, for a proper belief to take hold that management are really serious this time—and, of course, that will only happen if management really are. Management's commitment to constant improvement is a critical factor for securing the enthusiastic interest and involvement of employees at all levels, and for enabling them to contribute more. Such commitment can only be acquired by people in management taking the trouble to learn and understand deeply the new (to them) philosophy, and then setting a good example through their consistency of purpose constantly filtering down through the organisation to feed and nurture a constancy of purpose throughout. That requires action—action of a different nature from that which has been seen before; the right type of action will be evidence of constancy of purpose.

2. THE NEW PHILOSOPHY

Adopt the new philosophy. We are in a new economic age, created in Japan. We can no longer live with commonly-accepted levels of delays, mistakes, defective materials, and defective workmanship. Transformation of Western management style is necessary to halt the continued decline of industry.

It *is* a whole new philosophy. It is not merely just a few guidelines, ideas, rules, or techniques which you can tack on to the end of whatever you do now. It involves a thorough, radical rethink—more radical than you can imagine. It is likely to involve a complete reversal of attitude toward some strategies, modes of behaviour, and beliefs to which you have become accustomed and conditioned over the years. We are talking of deep, fundamental change—change as radical as that involved in conversion from the theory of a flat earth. If you do not accept the fact that we are talking about so fundamental a change, then it will not happen. In any case, it will not happen overnight. But there must be a constant, consistent movement in the right direction: every day there must be a move closer to total involvement in ever-improving quality of all systems, processes, and activities within the company.

3. CEASE DEPENDENCE ON MASS INSPECTION

Eliminate the need for mass inspection as a way to achieve quality by building quality into the product in the first place. Require statistical evidence of built-in quality in both manufacturing and purchasing functions.

If your initial reaction to this requirement from Dr. Deming is to laugh, then that only demonstrates how far away your standards are from those which he demands—and which are being achieved by those who have accepted his message. We have become so used to poor quality in supplies, systems, and service, and to such a high level of mistakes, errors, and defects, that we may have come to accept as a "fact of life" that this is the way things are and must forever be. But an undeniable result of reaching consistent high standards (such consistency being backed by statistical evidence and methods of process control) is that the expensive, nonproductive activity of mass inspection indeed becomes irrelevant, since scrap at source is eradicated. Considerable cost-savings will result, not only from the reduced need for inspection, but by the security of working with reliable, dependable, consistent, high-quality materials and processes, and all that that makes possible.

Just think what your resulting high-quality, competitive end-products and services will then do for your reputation, both with your existing and your potential customers.

4. END LOWEST–TENDER CONTRACTS

End the practice of awarding business solely on the basis of price tag. Instead, require meaningful measures of quality along with price. Reduce the number of suppliers for the same item by eliminating those that do not qualify with statistical evidence of quality. Move toward a single supplier[6] for any one item, on a long-term relationship of loyalty and trust. The aim is to minimise *total* cost, not merely initial cost. Purchasing managers have a new job, and must learn it.

This is fundamentally connected with the previous Point. The necessity for inspection of the input from our suppliers can only be ended if we can trust those suppliers to have the same high standards as ourselves. This implies a positive, cooperative, long-running relationship with a reduced number of chosen suppliers who can and will fulfil our needs. The savings obtainable from such a relationship with reliable suppliers, and the resulting trustworthy materials and service, outstrip to a dramatic degree the "savings" attainable by merely going for the lowest price. In any case, the lowest-tender approach to procurement forces suppliers into short-termism, preventing them from considering longer-term interests. The costs incurred within our operation, and probably subsequent to it, as the result of using cheap, unreliable input are likely to be enormous, incalculable. At best, there will be substantial rework, delays, and irregular throughput within our operation; at worst, the bad material may slip through our operation, leaving it for our customer to find. And, if our customer suffers, be sure that he will make us suffer as a consequence, and rightly so.

---

[6] The terms "single supplier" and "single source" are used interchangeably in this book. Strictly speaking, they are of course not the same, for one supplier may use several sources, either internally or externally or both. The prime purpose of Point 4 is to reduce variation (see Chapter 22), and so the more appropriate sense is that of "single source."

5. IMPROVE EVERY PROCESS

Improve constantly and forever the system of planning, production, and service, in order to improve every process and activity in the company, to improve quality and productivity, and thus to constantly decrease costs. It is management's job to work continually on the system (design, incoming supplies, maintenance, improvement of equipment, supervision, training, retraining, etc.).

There is at present far too great a tendency to "hope for the best," to "turn a blind eye," and to "let things ride" regarding potential problems —only paying attention to them when they become obviously serious and may well have already caused the organisation some considerable harm. Far better to seek them out early, to "nip them in the bud," before they cause real trouble. This is a basic difference between intelligent management and crisis management. Never be content; even when some problems have been sorted out, and some improvement has thereby been obtained, it is in the nature of things that further improvement is always possible, but it will only be achieved if further problems in processes are identified and solved. By "problems" we imply both special and common causes of variation (see Chapter 4). That is, we must strive to make unstable processes stable, *and* to make stable but incapable processes capable, *and* to make capable processes yet more capable (see Chapters 11 and 12).[7] Problems are opportunities for improvement. And if you do not find out problems, be sure they will find *you* out.

6. INSTITUTE TRAINING

Institute modern methods of training for everybody's job, including management, to make better use of every employee. New skills are required to keep up with changes in materials, methods, product design, machinery, techniques, and service.

How can anybody, staff or management, do their job properly if they do not know what their job is? Learning needs to be as much a part of the

---

[7] I am indebted to Kieron Dey for this clarification.

job as "producing." Consolidated improvement is a product of learning. But training is short-sightedly regarded as "nonproductive" by many managements, and is thus one of the first things to go when finances become tight. How wrong! Think how little proper training costs, as a proportion of the total costs involved with an employee over the months and years he may be working for the company. It is minute in comparison with the potential advantage to the company of that worker understanding his job, i.e. knowing how to do it properly and to the company's best advantage, and thus at least standing a chance of doing it that way. And this does not even include the unquantifiable gain to the company of the worker gaining satisfaction and pleasure from doing a good job, and thus *wanting* to continue so doing and improving yet further.

7. INSTITUTE LEADERSHIP OF PEOPLE

Adopt and institute leadership aimed at helping people to do a better job. The responsibility of managers and supervisors must be changed from sheer numbers to quality. Improvement of quality will automatically improve productivity. Management must ensure that immediate action is taken on reports of inherited defects, maintenance requirements, poor tools, fuzzy operational definitions (see Chapter 7), and other conditions detrimental to quality.

If foremen or supervisors have to spend their time chasing the people for whom they are responsible, and browbeating them to "do a proper job" or to keep up to schedule, that in itself is clear comment on the low standard of the operation concerned. Management will be deluding themselves that it is the "workers' attitude" which is the cause of poor quality. Such delusion in itself constitutes one of the most solid obstacles to managers getting even to first base with the Deming approach. They can neither visualise nor believe in a world where workers are involved in and committed to continual improvement—without bribery or coercion, that is. An environment must be created where workers have genuine interest in their work, and are helped to do it well. And these are complementary activities: if they are interested, then they will want to do it well and accept help to enable that; and if it is made possible for them to do it well, then their interest will increase—and so the cycle continues. Far too often, one sees the opposite kind of cycle: the vicious circle. Conditions force a worker to do a bad job; so he loses some of the interest that he has, which results in him doing a yet poorer job, which lessens his interest still further, and so on.

8. DRIVE OUT FEAR

Encourage effective two-way communication and other means to drive out fear throughout the organisation so that everybody may work effectively and more productively for the company.

Anybody working in fear of his superiors cannot be working in true cooperation with them. The best that can be hoped for in such circumstances is to get people working in resentful acquiescence—maybe that is all that some superiors desire. However, this will never result in much progress. Successful joint working relationships achieve so much more than isolated individual efforts—but they will not be successful without the nourishment of mutual trust, confidence, and respect. Those working in fear try to withdraw from the attention of those of whom they are afraid. And how can you expect to get anything like their true potential from people whose main aim is not to be noticed? The following Point concerns the breaking down of barriers between departments. It is just as important to break down barriers between staff and their supervisors, between those supervisors and middle management, between middle and senior management, and between senior management and the Chief Executive. In a climate that is heavy with fear, top management will be out of touch with reality. They will be fed with what they want to hear. Bad news will be withheld, delayed, watered down, massaged. Mistakes and errors will be buried with much energy and creativity—energy and creativity which could be put to better use. Thus fear severely impairs the organisation's opportunities for learning and for improvement.

9. BREAK DOWN BARRIERS

Break down barriers between departments and staff areas. People in different areas, such as Research, Design, Sales, Administration, and Production, must work in teams to tackle problems that may be encountered with products or service.

Most companies are organised functionally, but they need to operate cross-functionally. Different sections of an organisation have their own interests, their own traditions, their own values, their own "sacred cows," often their own languages. So they may well feel the need to fight against their fellow-employees if their interests appear to be in conflict. If the employees must fight somebody, better that they fight the competition rather than themselves. Often the differences are more apparent than real. And frequently a minor change in one department can afford considerable help to another—often with the resulting desire to "return the compliment." But such will only happen if the departments concerned have real understanding of each others' difficulties, and if the management environment encourages cooperation rather than generating inner conflict. A common language of elementary statistical methods and charting techniques is helpful in enabling people to gain an understanding of each others' jobs and problems, and how they may be helped.

10. ELIMINATE EXHORTATIONS

Eliminate the use of slogans, posters, and exhortations for the work-force, demanding Zero Defects and new levels of productivity, without providing methods. Such exhortations only create adversarial relationships; the bulk of the causes of low quality and low productivity belong to the system (see Chapter 4), and thus lie beyond the power of the work-force.

"Do it right the first time"; "Zero Defects is our aim"; "Increase output by 10%"; and countless others. Exhortation to produce better quality implies that management believe it is within the power of their operators so to do. This is bitterly resented by those operators who daily grapple with the output from incompetent systems. How can anybody do it right the first time if he is given neither the time nor the materials nor the equipment to make that feasible? How can he produce zero defects if what he gets to work on is already flawed? And his already-low job satisfaction will drop even more if he is exhorted to produce greater quantities which he knows will, under the prevailing detrimental conditions, lower the standards of what he is producing still further, however hard he tries to prevent it. Make *reasonable* requests, and provide what is necessary for them to be met, and you may well get *better* than you ask for. Make unreasonable requests, and you will get even less than you would have otherwise obtained from an increasingly demoralised worker.

11. ELIMINATE ARBITRARY NUMERICAL TARGETS

Eliminate work standards that prescribe quotas for the workforce and numerical goals for people in management. Substitute aid and helpful leadership in order to achieve continual improvement of quality and productivity.

If simplistic, imposed functional targets are used to "drive" the company, hitting those targets (or appearing to) becomes more important than satisfying the customer, let alone delighting him. Targets can never be "right," except maybe by very occasional accident. If a target is lower than what turns out to be reasonably achievable, the automatic reaction is for workers to slow down as that target is approached, and to take a holiday once it has been reached—and why shouldn't they? If the target is unreasonable, then either it will not be attained (resulting in criticism, loss of bonus, demoralisation—all at no fault of the workers), or it will be attained through cutting corners, lowering standards, ignoring safety requirements, etc.: the right numbers may be attainable, but at what cost in quality, with all the ramifications that that may have further down the line or, worse still, at the customer's? In either case, workers' respect for their management's ability to manage will justifiably take another dive.

12. PERMIT PRIDE OF WORKMANSHIP

Remove the barriers that rob hourly workers, and people in management, of their right to pride of workmanship. This implies, *inter alia,* abolition of the annual merit rating (appraisal of performance) and of Management by Objective. Again, the responsibility of managers, supervisors, foremen must be changed from sheer numbers to quality.

So many barriers to pride of workmanship exist, several of which have been touched on already. How can a worker be proud of what he is doing if he is being forced to produce shoddy goods because of poor materials, poor tools, unreasonable quantities of throughput being demanded? How can he be proud of what he is doing if he can see ways of improvement but knows it is pointless to try to discuss them with his superiors—so he reluctantly carries on in the same old way, which he knows to be a bad way? How can a manager be proud of what he does if the effect is to reduce quality and make his workers even less happy in their work? How can he be proud of what he does if there is neither time nor encouragement to try to improve morale and productivity by instigating improvements to processes and methods in order to raise quality? The value of what a worker, of whatever rank, produces will be immeasurably higher if he is enabled and encouraged to take pride in his work, compared with what he does if he is merely serving time. And what has merit rating to do with all this? You may well ask.

13. ENCOURAGE EDUCATION

Institute a vigorous programme of education, and encourage self-improvement for everyone. What an organisation needs is not just good people; it needs people that are improving with education. Advances in competitive position will have their roots in knowledge.

In older versions of the 14 Points, both Points 6 and 13 referred to education and training. Now Point 6 is devoted solely to training, and Point 13 to education plus a new concept: self-improvement. The distinction is important. Of course, training for the current job, in the way that that job is currently perceived, is necessary. But broader education is a great investment for the future. Things change fast in the modern world. Naturally, there is little point in change for change's sake. But, without being aware of change, and the potential advantage that it might bring, how can we or the company have any chance of working effectively to benefit from it? How can things improve without change? And how can change occur without knowledge of it or the education to create it? The recent addition to this Point concerning self-improvement represents a new challenge, as great if not greater than those requiring single-sourcing of suppliers and elimination of mass inspection presented years ago. Can you envisage the added potential of a work-force which actively seeks, without persuasion, instruction, or monetary reward, to increase its education, perhaps even at considerable personal expense?

14. TOP MANAGEMENT COMMITMENT AND ACTION

Clearly define top management's permanent commitment to ever-improving quality and productivity, and their obligation to implement all of these principles. Indeed, it is not enough that top management commit themselves for life to quality and productivity. They must know what it is that they are committed to—i.e. what they must do. Create a structure in top management that will push every day on the preceding 13 Points, and take action in order to accomplish the transformation. Support is not enough: action is required.

It all begins, and can end, here. Without full top-management belief, understanding, conversion, and action, progress (if any) will be sporadic and temporary at best. Top management must lead the whole organisation in the drive for ever-improving quality of every activity in the company by providing proper encouragement, training, facilities, time —and by practising what they preach. In particular, they must accept that they also have much to learn, and be prepared to learn it. What, for example, is the point of training everybody from middle management downwards in statistical techniques if higher management cannot, or rather will not, understand the reports, results, analyses, and recommendations emanating from the use of these methods? Indeed, more important still, they should be using them on their own data. Of course, top managers are very busy people. And that is why it is so necessary to set up a positive and lasting structure within management with the sole task of encouraging and facilitating continuing and continual progress in the new direction. It is hard work—Deming has never claimed otherwise—and the need for commitment and faith will never have been greater. But the potential rewards for you and your company are huge.

Chapter 2 of *Out of the Crisis* introduces the 14 Points. Chapter 3 then lists and discusses a number of "Deadly Diseases" and "Obstacles to the Transformation." We conclude this chapter by briefly summarising these Diseases and

Obstacles. The words in these final pages of the chapter are mostly abstracted and sometimes slightly amended from *Out of the Crisis* and Deming's 1982 book: *Quality, Productivity, and Competitive Position.* Yet again, the messages expressed here specifically to an American readership regrettably have much wider relevance.

The 14 Points constitute a theory of management. Their application will transform Western style of management. We will now describe some of the Diseases which stand in the way of transformation: they each have deadly effects. Alas, the cure for some of the Diseases requires complete shake-up of Western style of management (fear of takeover, for example, and owners that are interested only in short-term profit).

The Deadly Diseases afflict practically all big American companies. An esteemed economist, Carolyn Emigh, has remarked that cure of the Deadly Diseases will require total reconstruction of American management. Acceptance of the Deadly Diseases is the American way. Can total change take place? It will, yes, in companies that survive. How short is the time-fuse?

THE DEADLY DISEASES

1. LACK OF CONSTANCY

The crippling disease in America is lack of constancy of purpose to stay in business by planning to provide product and service in the future that will help man to live better materially, and which will have a market, and provide jobs.

Continual improvement of processes will raise quality and dependability of product and of service, and will decrease costs. A stockholder that needs dividends to live on is more interested in future dividends, not just the size of the dividend today. Give him cause to expect there will be dividends three years from now, five years from now, eight years from now—as long as he wants.

2. SHORT-TERMISM

Short-term thinking defeats constancy of purpose to stay in business with long-term growth.

Banks could help long-range planning, and thus protect funds entrusted to them. In contrast, the following actual quotation from an American banker is typical: "Jim, this is not the time to talk about quality and the future. This is the time to cut expenses, close plants, cut your payroll." Pursuit of the quarterly dividend and short-term profits destroys constancy of purpose.

3. APPRAISAL OF PERFORMANCE

The effects of performance appraisal (personal review system, merit rating, evaluation of performance, annual review, system of reward, pay for performance, etc.) are devastating.

"Management by Objective" is a similar evil; "management by fear" would be an even more appropriate title. The results of such practices are as follows:

- They nourish short-term performance, rivalry, and politics; they annihilate long-term planning, build fear, demolish teamwork;

- They leave people bitter, others despondent and dejected, some even depressed, unfit for work for weeks after receipt of rating, unable to comprehend why they are inferior. They are unfair, as they ascribe to the people in a group differences that may well be caused totally by the system within which the group works.

See Chapter 30.

4. JOB-HOPPING

Mobility of management causes instability, results in decisions being made by people who do not know the business (i.e. *this* business) and thus blindly use experience gained elsewhere which may be totally irrelevant.

It takes time to become a full part of any organisation, to become familiar with its business, its problems, its people, its customers. Experience gained elsewhere is only valuable if the user also has sufficient understanding of his current situation to ensure its validity (see Chapter 16). Much of American industry is run on the quarterly dividend. A President is brought in by the Board of Directors for this very purpose. He accomplishes it, leaves a path of destruction, then moves on to destroy another company.

5. USE OF ONLY VISIBLE FIGURES

One cannot be successful on visible figures alone. Of course, visible figures are important: the bank account must be watched, and employees and vendors must be paid. But he who would run his company on visible figures alone will in time have neither company nor figures.

Anyone can jack up the figures at the end of a quarter: ship everything on hand, regardless of quality; or mark it shipped, and show it all as accounts receivable; cut down on research, education, training; let go of people who are engaged for quality. Recall Lloyd Nelson's observation that the most important figures for management are unknown and unknowable.

Deming also mentions two other Deadly Diseases which he says are "peculiar to industry in the United States." They are excessive medical costs and costs of liability. Unfortunately, there are signs that those infections are spreading further afield.

There are besides the "Deadly Diseases" a whole parade of Obstacles. Some are nearly as dangerous as the Diseases, though most of them should be easier to cure—N.B. *easier*, not *easy*.

THE OBSTACLES

HOPE FOR INSTANT PUDDING[8]

"Come, spend a day with us, and do for us what you did for Japan."

THE SUPPOSITION THAT SOLVING PROBLEMS, AUTOMATION, GADGETS, AND NEW MACHINERY WILL TRANSFORM INDUSTRY

Quality cannot be bought with money. See the Author's Foreword and Chapter 17 of this book.

---

[8] This term was first suggested to Dr. Deming by Jim Bakken, formerly the Director of Corporate Quality at Ford Motor Company.

SEARCH FOR EXAMPLES

Examples teach nothing unless they are studied with the aid of theory. Most people merely search for examples in order to copy them. See Chapter 16.

"OUR PROBLEMS ARE DIFFERENT"

Maybe. But the principles that will help to solve them are universal.

OBSOLESCENCE IN SCHOOLS[9]

Students in Schools of Business are taught that there is a profession of management: that they are ready to step into top jobs. This is a cruel hoax. The MBA teaches managers how to take over companies, but not how to run them.[10]

POOR TEACHING OF STATISTICAL METHODS
IN INDUSTRY

Confidence intervals, tests of significance, etc. at most say something about what we already have. To imply that they are useful for prediction and planning for improvement is deceptive and misleading.

USE OF MILITARY STANDARD 105D AND OTHER TABLES
FOR ACCEPTANCE

They imply that there is an acceptable level of errors and faults, thus denying the need for improvement. See Chapter 21.

"OUR QUALITY CONTROL DEPARTMENT
TAKES CARE OF ALL OUR PROBLEMS OF QUALITY"

Would that it could! See Chapters 21 and 22.

---

[9] i.e. Schools of Business

[10] Partly derived from an article by Edward A. Reynolds in *Standardization News* (Philadelphia), April 1983.

"OUR TROUBLES LIE ENTIRELY IN THE WORK-FORCE"

Pleasant dreams. The workers are handicapped by the system, and the system belongs to management. See Chapters 4 and 28.

FALSE STARTS

Related very much to the above "Hope for instant pudding" and "We installed quality control," to follow. Wholesale teaching of statistical techniques, Quality Circles, suggestion schemes, Employee Participation programmes, etc. are all attempted short-cuts. There are no short-cuts.

"WE INSTALLED QUALITY CONTROL"

What is important for quality is not techniques but knowledge. Techniques and machines can be installed: knowledge and understanding cannot. See Chapter 31.

THE UNMANNED COMPUTER

A computer can be a blessing. It can also be a curse. "The data are in the computer." And there they sit.

THE SUPPOSITION THAT IT IS ONLY NECESSARY TO MEET SPECIFICATIONS

Specifications cannot tell the whole story. Meeting specifications can only satisfy customers, no more. The supposition is a bar to improvement. See Chapter 11.

THE FALLACY OF ZERO DEFECTS

Companies go out of business even though they make product or service without blemish, without defects. See also the previous Obstacle.

INADEQUATE TESTING OF PROTOTYPES

A prototype is one-off, with measurements artificially close to nominal. Without knowledge of variation, there can be no prediction. See Chapters 16 and 18.

"ANYONE THAT COMES TO TRY TO HELP US MUST UNDERSTAND ALL ABOUT OUR BUSINESS"

Why? Competent men in the business know all there is to know about their work—except how to improve it. Improvement requires a new kind of knowledge. It is the people in the company who will have to bring about improvement, by fusing this new knowledge with that which they already have.

These lists of Diseases and Obstacles are by no means exhaustive. Many readers will be able to suggest additions which are particularly relevant to their own field or in their own country. For example, John Coss (of the PQCS Management Consultancy) has provided me with the following particularly British items (some of which may also apply to business in the U.S. and other nations):

- Reluctance to learn from others;

- Class divisions and antagonisms;

- Too many accountants, too few engineers and statisticians;

- Regarding education and training as cost, not investment; and

- Excessive regard for tradition.

Chapter 4

VARIATION AND THE CONTROL
OF PROCESSES

"If I had to reduce my message for management
to just a few words,
I'd say it all had to do with reducing variation."

Dr. Deming often speaks of "horrible examples" that he has seen. Here is one from my own recent experience in England.

During an afternoon in the middle of a conference, delegates were invited to go on factory visits. The company whose plant I visited is a household name, well-known both nationally and internationally.

I was told that the company had recently "installed SPC." That, in itself, was a danger-signal.[1] Control charts by the dozen were pointed out to me. I enquired how it was decided which processes, and which measurements from those processes, would be charted. I learned that suggestions were made through some kind of brainstorming, and then a short-list was generated

---

[1] As we saw in Chapter 3, the notion that "We installed quality control" is one of the Obstacles to Transformation. Don Wheeler, in the video *A Japanese Control Chart*, also speaks eloquently of SPC being a whole new way of thinking, not just a technique.

by a multi-voting process. I don't want to comment on that here; of more concern is what I was told next. After the short-list was generated, its components were carefully investigated in order to find out which of them "were suitable for SPC." I was intrigued as to what that meant. After experiencing some difficulty in communication, the interpretation which eventually got through to me was that "suitable for SPC" apparently meant the same thing as "in statistical control."

So what, enquired I, happened to those processes which were not "suitable for SPC"? The answer was something like the following: "Ah well, we all know we have lots of clapped-out machinery around here, totally incapable of meeting modern-day standards. We know it should be scrapped and replaced by up-to-date equipment. So, since we are heading toward the day when we are supposed to have SPC on all of our processes, we requisition new machines which will be suitable for SPC." Rather hesitantly, I enquired if they realised that, while a process was out of statistical control (or, rather, was "unsuitable for SPC"), *nobody* could accurately evaluate its capability. As I rather expected, the question met with a perplexed silence and some pitying looks: clearly, I understood nothing about the matter.

Eventually, I took one relatively senior person aside and suggested, as politely as I could, that it could just be that the company was wasting millions of pounds on unnecessary new equipment. The response was, to say the least, cool.

Sadly, although the company had "installed SPC," they knew nothing of the work of Walter Shewhart who, as we have seen in Chapter 2, spearheaded real understanding of the behaviour and causes of variation and invented the control chart as long ago as the 1920s.

I have mentioned this story to several people. Many have similar tales to tell. It is tragic to see so much in the way of good intentions, best efforts, and expenditure going to waste. The cause is, simply, poor understanding and poor teaching—and these two naturally support and reinforce one another, continually making things worse.[2] We shall return to this story later in the chapter, by which time we shall understand better where the company was going wrong.

Let us now learn more of Shewhart's early work, and how it came about. Shewhart had worked for some 18 months at Western Electric (before Deming took his vacation-jobs with them in 1925 and 1926) and had then moved to the recently-founded Bell Laboratories in New York. I will let Deming take up the story (as he told it to an audience in Versailles, France on 6 July 1989—see the BDA booklet: *Profound Knowledge*):

> "Part of Western Electric's business involved making equipment for telephone systems. The aim was, of course, reliability: to make things alike so that people could depend on them. Western Electric had the ambition to be able to advertise using the phrase: "as alike as two telephones." But they found that, the harder they tried to achieve consistency and uniformity, the worse were the effects. The more they tried to shrink variation, the larger it got. When any kind of error, mistake, or accident occurred, they went to work on it to try to correct it. It was a noble aim. There was only one little trouble. Things got worse.
>
> Eventually, the problem went to Walter Shewhart at the Bell Laboratories. Dr. Shewhart worked on the problem. He became aware of two kinds of mistakes:
>
> 1. Treating a fault, complaint, mistake, accident as if it came from a special cause when in fact there was nothing special at all, i.e. it

---

[2] This could be construed as an example of Rule 4 in the Funnel Experiment—see Chapter 5.

came from the system: from random variation due to common causes.[3]

2. Treating any of the above as if it came from common causes when in fact it was due to a special cause.

What difference does it make? *All the difference between failure and success.*

Dr. Shewhart decided that this was the root of Western Electric's problems. They were failing to understand the difference between common causes and special causes, and that mixing them up makes things worse. It is pretty important that we understand those two kinds of mistakes. Sure we don't like mistakes, complaints from customers, accidents; but if we weigh in at them without understanding, we only make things worse. This is easy to prove by mathematics."[4]

And we shall see a famous example in which this lack of understanding "makes things worse" later in this chapter.

So the purpose of Shewhart's work was to *improve quality* by *reducing variation*, exactly in the spirit of Deming's words at the start of this chapter. As we have seen in Chapter 2, Deming soon realised that Shewhart's ideas were relevant to a much wider range of processes than just those concerned with manufacturing operations. And, in the ensuing decades, developments of Shewhart's ideas have laid the foundations of Deming's whole management philosophy. In particular, basic arguments leading to his call for abolition of such common management practices as MBO, performance appraisal, and the use of arbitrary numerical

---

[3] The introduction to Shewhart's work in Chapter 2 provides a review of the essential nature of common and special causes of variation.

[4] Dr. Deming's thinking here is again illustrated by the Funnel Experiment (q.v.).

targets (Points 11 and 12) follow from the concept of *tampering* with stable systems (as illustrated by the automatic tool compensation example later in this chapter; see also Chapter 5).

Let us now examine what Shewhart meant when he spoke of *controlled* and *uncontrolled* variability and, analogously, what is meant by a process being *in* or *out* of statistical control. The essential ideas are not difficult—though, as we have implied, the consequences are far-reaching indeed.

Suppose that we are recording, regularly over time, some measurements from a process. These measurements might be the lengths of steel rods after a cutting operation, or the length of time required to service a machine, or your weight as measured on your bathroom scale each morning, or the percentage of defective (or "nonconforming") items in batches from a supplier, or measurements of Intelligence Quotient, or the times between sending out invoices and receiving the money, etc.. Figures 5a–5d show four records (run-charts) of this type. In each case, time is measured along the horizontal and, the higher the points, the larger are the sizes, lengths, times, or whatever else is being represented.

Figures 5a and 5b are typical of what we might expect from processes which are *in* statistical control; Figures 5c and 5d give pretty clear indications of processes being *out* of statistical control. All four graphs show variation in the measurement (for, without variation, a graph is just a horizontal line). The difference is that in Figures 5a and 5b the nature of the variation stays effectively the same throughout the time-period covered, whereas in Figures 5c and 5d there are some notable changes of behaviour in the variation as time progresses. What effect does this have in practice? Something very important indeed. In the case of Figures 5a and 5b we have a chance to predict what the process will produce in the future (with no certainty, of course— something may turn up to upset things). But with Figures 5c and 5d we can make no prediction, for the behaviour of the output from those processes is changing in literally unpredictable ways.

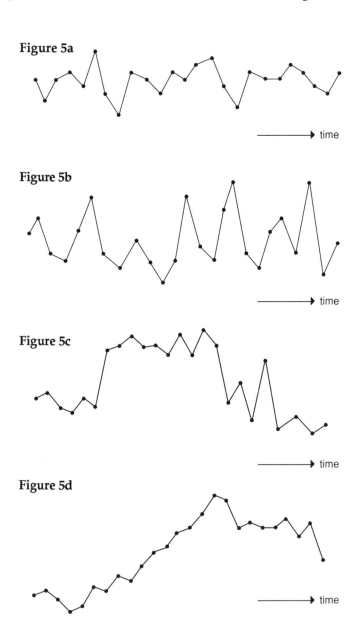

FIGURE 5
Four Run-Charts

A somewhat more formal interpretation of what is meant by a process being in or out of statistical control is given in a single page of diagrams from a Ford (of America) Motor Company manual,[5] and those diagrams form the basis of our Figures 6–9.

In Figure 6, we see an "at-a-glance" notion of a statistical distribution. The little boxes are positioned along the horizontal according to the values of "size" that we measure. A stack of such boxes representing a whole sample is called a *histogram.*

Suppose we record more and more data (and adjust the vertical scale in order to stop the histogram going off the top of the page!). Then, under certain important conditions discussed below, the picture will eventually settle down and change only imperceptibly as more data are added. It has become a picture of the measurement's statistical distribution. It is a picture of the behaviour of the variation in the measurements we are recording. The data and distribution in Figure 6 refer to the "size" of "pieces," but the context could just as well be any of the examples suggested above, or any of a host of other possibilities.

The vital words in Figure 6 are: "if the underlying process is stable"; this directly relates to the concept of the process being in statistical control. The point is that, if an outside influence impacts upon the process while the data are being collected (e.g., in the case of the above examples, a machine-setting is altered, a servicing engineer's quota is increased, you go on a diet, your supplier starts using inferior raw materials, etc.), then the numbers are of course no longer all coming from the same source, and so no single distribution could serve to represent them. (In fact, as we shall discuss later, the notion of stability as represented by a single fixed distribution is over-idealised from a practical viewpoint—we shall refer later to this artificial ideal situation as *exact* stability.)

As seen in Figure 7, distributions may differ in all sorts of

---

[5] *Continuous Process Control and Process Capability Improvement*

PIECES VARY FROM EACH OTHER

BUT, IF THE UNDERLYING PROCESS IS STABLE, THEY FORM A PATTERN WHICH
IS CALLED A STATISTICAL DISTRIBUTION

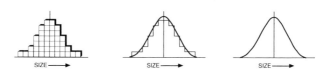

FIGURE 6
Building a Statistical Distribution from Data

(Adapted from Ford Motor Company, *Continuing Process Control and
Process Capability Improvement*, page 4a)

DISTRIBUTIONS CAN DIFFER IN:

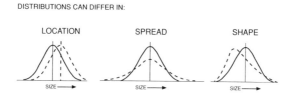

. . .OR ANY COMBINATION OF THESE

FIGURE 7
Types of Differences Between Distributions

(Adapted from Ford Motor Company, *Continuing Process Control and
Process Capability Improvement*, page 4a)

ways. "Location" refers to the average; "spread" refers to the scale of variability about the average, and "shape" indicates, for example, whether the data occur symmetrically either side of the average or if instead there is something of a pile-up on one side and a wider spread on the other.

In terms of statistical distributions, Figures 8 and 9 indicate respectively, again "at-a-glance," what is meant by a process being in or out of statistical control. The process is *in control* if the underlying distribution remains, in effect, unchanged as time proceeds; if the distribution changes substantially and unpredictably over time, then the process is said to be *out of control*.

As a scientist, Shewhart knew that there is always variation in anything that is being measured. The variation may be large, it may be imperceptibly small, or it may be between these two extremes; but it is always there.[6]

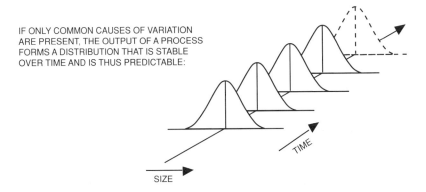

IF ONLY COMMON CAUSES OF VARIATION ARE PRESENT, THE OUTPUT OF A PROCESS FORMS A DISTRIBUTION THAT IS STABLE OVER TIME AND IS THUS PREDICTABLE:

TIME

SIZE

FIGURE 8
Distributional View of an *Exactly* Stable Process

(Adapted from Ford Motor Company, *Continuing Process Control and Process Capability Improvement*, page 4a)

---

[6] Exceptions can occur where data are obtained by *counting* rather than *measuring*. Hopefully, a manufacturer of dice can ensure that his dice have six faces! Even so, the weights, coloration, etc. will vary.

What inspired Shewhart's development of the statistical control of processes was his observation that the nature of the variability which he saw in manufacturing processes often differed from that which he saw in so-called "natural" processes, by which he meant such phenomena as molecular motions. On page 5 of *Understanding Statistical Process Control*, Donald J. Wheeler and David S. Chambers combine these two important observations as follows:

> "While *every* process displays variation, some processes display *controlled* variation, while others display *uncontrolled* variation."

In particular, Shewhart often found controlled (stable) variation, as exemplified in Figures 5a, 5b, and 8, in natural processes and uncontrolled (unstable) variation, as in Figures 5c, 5d, and 9, in manufacturing processes. The difference is clear. In the former case, we know what to expect in terms of variability: the process is *in statistical control*; in the latter case, we don't: the process is *out of statistical control*. We may predict the future, with some chance of success, when the process is in control; we cannot do so when the process is out of control.

Let us be clear what we mean by "predict" in this context. We do *not* imply that we expect to forecast in any exact sense what the future values from the process will be. Traditional statisticians sometimes talk of "point estimators" or "point predictors," which may give the impression that such exactness is feasible. But all that they are actually doing is producing some kind of *average* of what may be expected: we also need knowledge of the *variability* around that average in order to learn something realistic about the likely future values.

Let us look at three particular implications of what we have learned.

First, when the output from a process is affected by special causes, i.e. its behaviour changes unpredictably, it is impos-

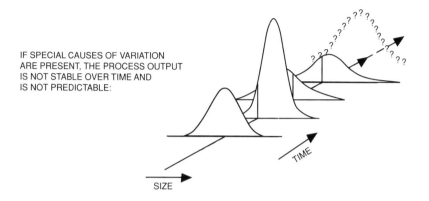

IF SPECIAL CAUSES OF VARIATION
ARE PRESENT, THE PROCESS OUTPUT
IS NOT STABLE OVER TIME AND
IS NOT PREDICTABLE:

TIME

SIZE

FIGURE 9
Distributional View of an Unstable Process

(Adapted from Ford Motor Company, *Continuing Process Control and
Process Capability Improvement*, page 4a)

sible to evaluate the effects of changes in design, training, purchasing policy, etc. which might be made to the process (or to the
system which contains the process) by management. While a
process is out of control, no one can predict its capability. This is
the point I tried in vain to make to that company which had
"installed SPC." (See Deming's discussion on "Capability of the
process," pages 339–340 in *Out of the Crisis*.)

Second, when special causes have been eliminated so that
only common causes remain, improvement then has to depend upon
management action. For such variation is due to the way that the
processes and systems have been designed and built—and only
management has authority to work on systems and processes. As
Myron Tribus, Director of the American Quality and Productivity
Institute, has often said:[7]

---

[7] See, for example, his papers: "Creating the Quality Company" and
"Creating the Quality Service Company."

> "The people work in a system.
> The job of the manager is
> to work on the system
> to improve it, continuously,
> with their help."

And, third, we come to Western Electric's problem with their telephone equipment: if we (in effect) misinterpret either type of variation as the other, and act accordingly, we not only fail to improve matters—we literally *make things worse.* Clearly, this phenomenon has to remain a mystery to those who lack understanding of the nature of variation.

These implications, and consequently the whole concept of the statistical control of processes, had a profound and lasting impact on Dr. Deming. Many aspects of his management philosophy emanate from considerations based on just these notions. As indicated earlier, quite apart from powerful humanitarian arguments, the two most controversial of his 14 Points—those seeking the abolition of arbitrary targets and performance appraisals—have their foundations right here. First, if a target is beyond the capability of a system to produce, the only way to achieve it is to distort the system in a way which causes trouble elsewhere: secondly, the amount of common cause variation in a person's appraised performance is such that it almost always effectively obscures the person's actual contribution.[8]

Dr. Joseph Juran[9] estimated many years ago that no more than 15% of the problems (or opportunities for improvement) in an organisation are due to special causes (i.e. possibly, but not exclusively, the province of the workers), leaving management with the responsibility for at least 85% of potential improve-

---

[8] Deming also produces many further arguments concerning these two Points—see Chapters 29 and 30.
[9] Joseph M. Juran is another famous American writer and speaker on quality; also like Deming, Dr. Juran is well-known for his work in Japan, where he first visited in 1954.

ment through changes on the system within which their employees are obliged to labour. After quoting these figures for many years, Dr. Deming revised them in 1985 to 6% and 94% respectively.[10]

It is very often the case that workers, if they are asked, can identify the special causes that result in difficulties which prevent the system from working to its full capability—after all, they are the ones who have to suffer them directly. But only management can change the actual system within which they work and which contains by far the bulk of the obstacles to improved quality, reliability, and productivity; though, as Tribus points out, they will probably still need the help of the workers to find out what they need to do. However, in no way can workers change the system by themselves. As Deming puts it, a "worker, when he has reached statistical control, has put into the process all that he has to offer" (*Out of the Crisis*, page 405; see also Chapter 24 in this book).

The promised illustration of the harm that can be caused by misinterpreting one type of variation as the other comes from Ford Motor Company (see pages 29–31 of Bill Scherkenbach's[11] book: *The Deming Route to Quality and Productivity*).

Input shafts for a transmission were turned in a machine equipped with an automatic compensation device. If the diameter of a shaft was measured as being too large, the compensation device reduced the machine setting by an amount equal to the discrepancy; similarly, if the diameter was too small, the machine setting was increased by that amount. Sounds sensible? Of course.

Figure 10 is a histogram of the diameters of 50 shafts consecutively manufactured by this process. The suggestion was

---

[10] I have recently heard it rumoured that the figures have been updated again—this time to 2% and 98%!
[11] William W. Scherkenbach, who first studied under Dr. Deming at the New York University Graduate School of Business in 1972, was for several years Director of Statistical Methods at the (American) Ford Motor Company.

made by a statistician that a similar set of 50 readings should be taken with the compensation device turned off. Figure 11 was the result: reduced variation, i.e. better quality. How could this be?

The answer is that the production process was already in statistical control. Without the compensation device operating, the process was already exhibiting the lowest variability of which it was capable: only common causes were present. Reduction of that variability could therefore only be achieved by improvement of the process itself. The compensation device was not improvement of the process. It was simply *tampering* with a process which was already stable. ("Tampering" is Dr. Deming's own term for this effect.)

Since the variability was already down to its minimum possible level *without* the compensation device, the tampering by the device constituted an "outside influence," as we referred to it earlier. The only possible effect of such an outside influence is to *increase* the variation—exactly the opposite effect from that desired. Of course, had there been special causes present, the compensation device might well have helped smooth out their effect. But, in the absence of special causes, it could only *harm* the output. It can be proved that, in such a case, the compensation device increases the variability by just over 40%.

This example demonstrates the necessity for management to understand variation in the way that Shewhart explained it. And the above form of compensation is actually one of the *least* harmful types of tampering! Other well-intentioned attempts to improve matters can make things worse, not just by a factor such as 40%, but by whole orders of magnitude (see Chapter 5).

Furthermore, the Ford example took place in the relatively simple context of a manufacturing process. Deming has been known to say that the first control charts to be drawn in any organisation should be not on shop-floor processes but on the data which land on the Chief Executive's desk—such as the figures on budgets, forecasts, absenteeism, and accidents. Are these pro-

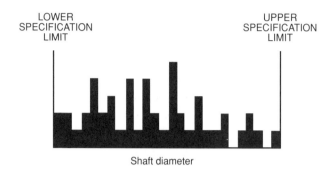

Shaft diameter

FIGURE 10

Histogram of Data: Automatic Tool Compensation Turned On:
Overcontrol **or** *Tampering*

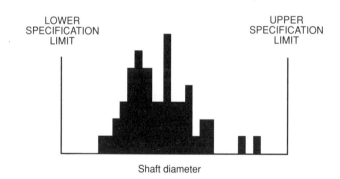

Shaft diameter

FIGURE 11

Histogram of Data: Automatic Tool Compensation Turned Off

(Adapted from William W. Scherkenbach,
The Deming Route to Quality and Productivity, page 25)

cesses "in control"? If so, are they being improved, or are they merely being tampered with, with results similar to those seen above or, as just suggested, many times worse?

This example of tampering shows the harm which can be caused by interpreting common cause variation as special cause variation—*the first kind of mistake.* Our opening "horrible example" made costly judgments concerning the capability of processes (which only relates to their common cause variability) when they were out of control—a case of *the second kind of mistake.*

Most of the examples which Deming mentions in this area concern the first kind of mistake: something undesirable (a fire, an accident, a complaint) happens, and the almost automatic reaction is to respond to it in isolation, as a one-off occurrence. The implication is that the *system* never does anything wrong. Would that that were so, for special causes are usually so much easier to identify and work on than are common causes. In fact, as Juran's 85%–15% and Deming's 94%–6% figures imply, unfortunately it is the system which causes the large majority of undesirable happenings. So treating them as special causes is tampering, with harmful results such as we have just seen. As in the Ford example, reaction to specific occurrences then *increases* overall variability, thus worsening quality, reliability, predictability of what will happen in time to come. This is a hard message for the newcomer—one of many to be encountered as he delves into Deming's work. But let's not argue about the *results.* Study the *theory;* for if the theory is unarguable, and the logic leading from the theory to the results is valid, how can the results be wrong (see Chapter 16)?

Some of the relevant illustrations in *Out of the Crisis* concern defective items on a production line (pages 54–55), road accidents (pages 316–317), fires (pages 324–325), colour-matching (page 329), firing torpedos (page 330), calibration of instruments (pages 330–331), faulty tubes for nuclear reactors (page 358), faulty tires (page 359), weights of copper ingots (page 359–360), and performance of managers at freight terminals (pages 361–362).

Now, let's be clear: we are not saying that direct action should never be taken following an accident, or a complaint, etc.. Some action will, of course, be necessary in any case—legal proceedings, apologies, replacement, etc.; that is not the kind of action being questioned. What we are questioning is the appropriate action for preventing or reducing the chance of trouble in the future. For that we need guidance on whether the occurrence indicated a special cause (necessitating direct action) or whether it was part of the system (in which case direct action is tampering, and more comprehensive improvement of the system as a whole is called for). How do we choose?

Shewhart developed a technical tool to help us differentiate between the two situations: the *control chart*. It is not our intention here to give full technical details on the construction and use of control charts—they are available from many other sources—see, for example, the late Kaoru Ishikawa's *Guide to Quality Control* and Donald J. Wheeler and David S. Chamber's *Understanding Statistical Process Control*. But here is an outline of the principles involved.

We are charting over time a measurement, or an average or range[12] of a set of measurements, or a count of defects or defectives; this is a *run-chart* or a *time-series*. Three horizontal lines are drawn on the chart: the Central Line, and the Upper and Lower Control Limits. The Central Line represents some kind of average[13] of the plotted points. The control limits are placed at 3 "standard deviations" of the plotted points either side of the Central Line. The standard deviation, often referred to by the Greek letter σ (sigma), is the most-recognised statistical measure of variability. Data widely spread around their average have a high standard deviation; data tightly clustered around their

---

[12] range = the difference between maximum and minimum values
[13] usually the mean, but sometimes the median

average value have a low standard deviation.

The formula by which σ is computed varies according to the type of data being charted; whenever possible, it is an attempt to reflect the standard deviation of the common cause variability. Shewhart's rule was to take action appropriate to a special cause if a plotted point lies beyond either control limit. Nelson's tests suggest also taking such action in response to other signals, such as 9 successive points on one side of the Central Line or 6 points in rising or falling sequence. Control charts for the data plotted in Figures 5a–5d are shown in Figures 12a–12d;[14] they bear out the informal judgments made earlier concerning which of these processes are, and are not, in control.

There is no claim that either Shewhart's rule or any of Nelson's tests *always* gives correct guidance. Remember the two kinds of mistakes. The gentler the rule for choosing direct action appropriate for a special cause, the more often we shall make Mistake Number 2 but the less often we shall make Mistake Number 1. The more eager the rule is to signal a special cause, the more often we shall make Mistake Number 1 but the less often we shall make Mistake Number 2. The aim is to make these mistakes *with minimum overall economic loss*. There is no exact solution; it is a matter of give and take:

---

[14] For the purpose of constructing these charts, the data in Figures 5a–5d have all been interpreted as individual measurements (as opposed to, say, means or numbers of defects, etc.). Consequently, the Central Line represents, in each case, the mean of the 25 observations, and the control limits are placed at a distance of 3.14 x the median moving range either side of the Central Line. (Moving ranges are the sizes of the differences between consecutive observations, and this method of calculating control limits for individual observations can be particularly effective for measuring common cause variability even in the presence of special causes.) A control chart for a different kind of data, along with appropriate calculations, will be seen in Chapter 6.

Figure 12a

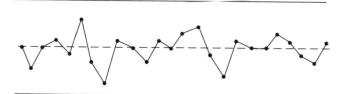

Figure 12b

Figure 12c

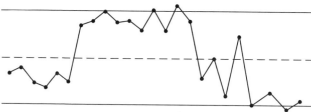

Figure 12d

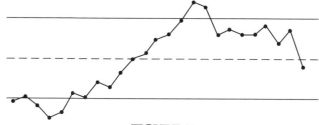

FIGURE 12
Four Control Charts

> *"What shall we do?*
> *How shall we do it?*
> *Dr. Shewhart helped us*
> *with these important questions,*
> *and this was a great contribution*
> *to man's process of thought and ability to manage."*

Shewhart's choice of 3σ as the distance between the Central Line and the control limits, as opposed to any other multiple of σ, did not stem from any specific mathematical computations—his reason was simply that it "seems to be an acceptable economic value" (see page 277 of his 1931 book). This healthy, pragmatic approach differs markedly from more strictly mathematical approaches to setting control limits which we shall discuss at the end of this chapter.

Even worse are ideas by which the control limits are not even computed from data taken from the process. Deming reports two examples in *Out of the Crisis*, pages 356–358 (one even coming from a company which had won the Deming Prize in Japan) where the lines had been placed on charts "by judgment" or by "the manufacturer's claim." This section of *Out of the Crisis* is headed: "Examples of Costly Misunderstanding." Elsewhere in the same chapter he includes the not-uncommon error of drawing specification limits on charts instead of calculated control limits. The purpose of control limits is to indicate what the process is doing and what it can do. Of course, we have to consider what the customer says he needs, but to use specification limits rather than control limits on control charts only causes confusion:

> *"If you use specification limits as control limits,*
> *you'll be tampering,*
> *making things worse."*

So, to reiterate, the purpose of Shewhart's work was to

guide us to the type of action which is appropriate for trying to improve the functioning of a process. Should we react to specific, isolated results from the process (which is only sensible if the process is out of control) or should we instead be going for change to the process itself, guided by cumulated evidence from its output (only sensible if the process is in control)?

Process improvement needs to be carried out in three chronological phases:

Phase 1:   Stabilisation of the process (i.e. getting it into control) by the identification and elimination of special causes.

Phase 2:   Active improvement efforts on the process itself, i.e. tackling common causes; and

Phase 3:   Monitoring the process to ensure the improvements are maintained. From what might be called a pure Deming point of view, this version of Phase 3 is never reached, for it contradicts the aim of continual improvement. So we should incorporate into Phase 3 the need to implement additional improvements as the opportunity arises.

It should be pointed out here that some approaches to quality improvement concentrate unduly on Phase 1. That is just as bad as ignoring it (as was the case in the story at the start of this chapter). "Problem-solving" may fall into this trap. Some speak of the search for, and eradication of, special causes as "fighting fires." If fires are alight, they should certainly be stamped out. The point is however, that even if a fire in a building is successfully extinguished, this act does not improve the building—it simply stops it getting worse than it was before. Phase 1 merely gets the process to where it should have been already—doing what it was always capable of doing. Only then can *improvement* of the process commence.

Control charts have an important role to play in each of the three Phases. Points beyond control limits (plus other agreed signals) indicate when a search for special causes should be initiated. The control chart is therefore the prime diagnostic tool in Phase 1. All sorts of statistical tools can aid Phase 2, including Pareto analysis, Ishikawa diagrams, flow-charts of various kinds, etc. (see Ishikawa's book and Peter Scholtes' *The Team Handbook*), and recalculated control limits will indicate what kind of success (in terms of reduced variation) has been achieved. The control chart will also, as always, show when any further special causes should be attended to. In the first-sentence-only version of Phase 3, the purpose of the control chart is merely to diagnose the onset of any special causes which upset the state of stability reached at the end of Phase 2. In the fuller version, as in Phase 2, we may recalculate control limits after further changes to the process in order to assess their effects.

Shewhart's reasoning and approach to the construction of control charts were, as we have seen, practical, sensible, and positive. In order to be so, he deliberately avoided overdoing mathematical detail. Unfortunately, some years after Shewhart's first publications in the area, some mathematical statisticians (mostly British, it appears) seized upon Shewhart's ideas, filled in what they perceived to be the mathematical gaps, and hence fell into the trap (which Shewhart had been so careful to avoid) of *reducing* the usefulness of the techniques.

The problem is that, in order to enable the development of a sophisticated mathematical argument, it is usually the case that assumptions need to be made which are unduly restrictive in terms of the real world. Control charting is no exception; indeed, in this case, the assumptions required for the mathematics beg more than the important questions which Shewhart set out to answer. Even more unfortunately, this weakened version (though it is often perceived as strengthened through its mathematical exactness) spread and became better-known than Shewhart's own work, certainly in Britain and Europe.

Exact mathematical methods are both easier to teach and more impressive to the unwary. But they have seriously reduced the potential of what may be accomplished through the use of SPC. Without doubt, the company which "had installed SPC" had been trained by exponents of what I shall refer to as the probability approach. Who can blame the trainers? It was probably all they knew; they were only doing their best.

What is the "probability approach"? In the most usual version, control limits are calculated so that, *supposedly*, while the process is in control, there is a 1 in 1000 chance that a typical plotted point will lie above the Upper Control Limit and a 1 in 1000 chance that it will lie below the Lower Control Limit. Other versions include *two* pairs of control limits: those just mentioned are then referred to as Action Limits while an inner set, corresponding to probabilities of 1 in 40 rather than 1 in 1000, are called Warning Limits.

It is instructive to read pages 275–277 in Shewhart's 1931 book which, as you will recall, is where he first suggests the use of 3σ limits. He seems to toy briefly with the idea of using the probability approach. Let us paraphrase part of those three pages. He points out that, *if* a process were *exactly* stable and *if* we knew the details of its underlying (fixed) statistical distribution, we *could* then work in terms of probability limits. But then he faces reality and admits that, in practice, we would never know the appropriate type of statistical distribution. The mathematical statisticians usually plump, almost as if it's a foregone conclusion, for their favourite distribution, known as the *normal* distribution. Shewhart specifically disposes of that on page 12 of his 1939 book. In the 1931 book he said in effect that, even if the process were exactly stable and if the normal distribution were appropriate (neither of which we would ever know), we would still never know the value of its mean. And, even if we did, we would never know the value of its standard deviation. We could only estimate these from the data. The probability calculations depend on knowledge of *all* of these things.

In *Out of the Crisis* (see below), Deming throws an even larger spanner into the works by pointing out that, in practice as opposed to theory or hypothesis, *exactly stable processes never exist*. That is, real processes are never entirely free of special causes, in the sense defined as in the Ford exposition: i.e. Figure 8 never exists. This identifies a vast gulf between the common mathematical assumptions and the real world. What does this imply? Surely not that we are to spend all of our time looking for special causes and none on improving processes, i.e. working on common causes. No, of course not. What we need is guidance as to *when special causes are troublesome enough to warrant attention*. This way of thinking is extremely well reflected in the very definitions of controlled and uncontrolled variability which Deming used in Japan in 1950, though they do not appear in his current writings:

Controlled variability

> It will not be profitable to try to determine the cause of individual variations when the variability is controlled.

Uncontrolled variability

> It will be profitable to try to determine and remove the cause of uncontrolled variability.

Shewhart's control chart, with its 3σ limits, provides guidance for choosing between the two situations. His reason for selecting the whole number 3, rather than anything more fancy, was, as we have seen, simply that it "seems to be an acceptable economic value." No calculations from the normal, or any other, distribution were involved. Deming is adamant that:

> *"It is not a matter of probability.*
> *It is nothing to do with how many errors we make*
> *on average in 500 trials or 1000 trials.*
> *No, no, no—it can't be done that way."*

Pages 334–335 of *Out of the Crisis* shed further light on the matter:

> "It would ... be wrong to attach any particular figure to the probability that a statistical signal for detection of a special cause could be wrong, or that the chart could fail to send a signal when a special cause exists. The reason is that no process, except in artificial demonstrations by use of random numbers, is steady, unwavering. It is true that some books on the statistical control of quality and many training manuals for teaching control charts show a graph of the normal curve and proportions of area thereunder. *Such tables and charts are misleading and derail effective study and use of control charts.*"

(my italics). Deming's claim that we never get an exactly stable process in practice indicates the artificiality of the probability approach. It also shows the Ford description of the meaning of processes being in and out of control to be overly-simplistic.

However, the most serious drawback of the probability approach, from the viewpoint of how it has affected the practice of using control charts, is that it has given many practitioners a very limited view of what control charts are for. The company from whence came the "horrible example" at the beginning of this chapter was a clear case in point. Actually, those people seemed to have no notion of the importance of what we recently called Phase 1 in the improvement of a process. They certainly had no notion that the control chart could be used in that Phase.

Of course they hadn't—if a process is out of control (as it usually is when we first look at it), there most certainly is no distribution and no probability. The probability approach *starts* at Phase 2— and a very idealised version of Phase 2 at that.

Even companies that have, by some means or other, gained a better level of understanding still find themselves influenced by some of the probability approach's implications. For example, a company comes to mind for which I have great respect: I know several of their people very well. Yet, when I looked at this company's SPC manual, I found:

> "Statistical Process Control is a means by which to achieve the prevention of defects by highlighting situations when the output of the process is drifting outside acceptable limits ... "

and

> "Even with capable equipment, substandard work will be produced with
>
> • inappropriate operator adjustment
> • a broken or worn tool"
>
> (etc., etc.)

and

> "The use of Process Control Charts provides an early warning of these problems and will often anticipate them so that they can be avoided."

Now, there wouldn't seem to be anything inherently bad about these uses of control charts. These may well be useful things to do. However, there are hints here of both "tampering" and thinking of quality purely in terms of meeting requirements and specifications—and these are both to be avoided (see also Chapters 5 and 11 respectively). The point is that the above description reflects only a minor part of what control charts can do, whereas many people think that this kind of thing is *all* that they can do. They relegate the control chart to the role of a mere *monitoring* procedure. The process is presumed, by some means or other, to find itself in a satisfactory state, and the control chart's function is perceived to be as an early-warning device of when the process moves away from that satisfactory state.

The crucial difference between Shewhart's work and the wrongly-perceived purpose of SPC described above is that his work was developed in the context, and with the purpose, of process *improvement* as opposed to process *monitoring*, i.e. it could be described as concerned with helping to get the process into a "satisfactory state" which one might then be content to monitor. (However, remember our comment about a pure Deming approach never really getting to the third of our three Phases of improvement.)

As many readers will realise, this difference is far more important than some might at first appreciate. It gets right to the heart of the divide between the main approaches to the whole quality issue. On the one hand, we have approaches which regard quality merely in terms of conformance to requirements, meeting specifications, Zero Defects. On the other hand, we have Deming's demand for continual improvement—a never-ending fight to reduce variation. The probability approach can only cope with the former. Shewhart's own work was inspired by the need for the latter.

PART 2

SOME FUNDAMENTALS

SOME FUNDAMENTALS

Part 2 contains a variety of topics, ranging from some which already have good coverage in *Out of the Crisis*, through some which are mentioned only relatively briefly in *Out of the Crisis*, to others which are important to Dr. Deming's work, though are not usually treated as specific topics by him.

We begin with two famous experiments. These are the Funnel Experiment and the Experiment on Red Beads: supreme demonstrations on the nature of variation, and for learning about its crucial importance to management thinking, and for understanding why much conventional practice goes wrong. Both of these experiments are also described in Chapter 11 of *Out of the Crisis*. Chapter 9 in *Out of the Crisis* is concerned with operational definitions, the subject of our Chapter 7. This is the one case in which I have included a chapter specifically on the same subject as a chapter in *Out of the Crisis*, and this is of course in recognition of its paramount importance—both Deming and Shewhart have placed it on a par with the understanding of variation.

The words "process" and "system" are frequently used by Deming and others, but some listeners and readers may have a rather limited notion of what is implied by these words. Chapter 8 attempts to redress this balance. Chapter 9 introduces and describes the Deming Cycle (also known as the Shewhart, PDCA, and PDSA Cycle). It is one of those notions which makes such obvious good sense that we often forget to employ it. Yet it surely has to be fundamental to both improvement and innovation in quality.

Chapter 10 is a warning, or rather many warnings—warnings about using numbers. To some, the receipt of such warnings from statisticians may seem paradoxical. But I would submit that an indication of the worth of a statistician is the awareness which he has of the *limitations* of numbers. Lord Kelvin told us in 1883 that, when we cannot express something in numbers, our "knowledge is of a meagre and unsatisfactory kind," and that then we have scarcely "advanced to the state of science." However, as Myron Tribus has perceptively stated:[1]

> "The problem is that we are not dealing with a *science*. We are dealing with the *art* of management."

Finally, in this Part, we have two chapters on the Taguchi loss function. Chapter 11 discusses why the Taguchi loss function is a "better description of the world" (*Out of the Crisis*, page 141) than the yes-no criterion of conformance to specifications. As an obvious contribution, the Taguchi loss function is consistent with the philosophy of continual improvement, whereas conformance to specifications is not. Chapter 12 goes into a few technical investigations based on the Taguchi loss function. This is the one chapter of this book which could prove a little daunting to non-mathematical readers, who are therefore invited to regard it as optional!

---

[1] "Judging the Quality of an Organisation by Direct Observation"

Chapter 5

THE FUNNEL EXPERIMENT

As we have already noted several times, Deming tells us that good intentions, best efforts, hard work will not produce quality. Why not? Can there be anything bad about good intentions and hard work? Sadly, the answer is yes. We all know the saying that "the road to Hell is paved with good intentions." Deming's version is:

"We are being ruined by best efforts."

If the best efforts and hard work are directed at the wrong things, or at the right things but in the wrong ways, we can finish in a worse position than that in which we started. The harder a man struggles in quicksand, the faster he disappears from view.

Best efforts and hard work are, of course, often directed at putting right those things which are wrong. The Ford tool compensation example (Chapter 4) is a prime illustration of how best efforts can instead increase variation, and thus harm quality, rather than the desired and intended converse. That example is, as we shall see, a clear case of the use of Rule 2 in the Funnel Experiment which we shall now describe. Nobody is saying that best efforts and hard work are *inherently* wrong; but knowledge— Profound Knowledge—is needed to ensure that such virtues produce rewards rather than disappointments.

The Funnel Experiment, which emanates from a suggestion made to Dr. Deming by Lloyd Nelson in 1986, is a very simple physical model which shows how some best efforts to put things right can go disastrously wrong. And the experiment is no purely academic exercise. Evidence for that comes repeatedly from Dr. Deming's four-day seminars. Within an hour or two of being introduced to the experiment, delegates are often asked to describe practical situations from their own experience which are, as they are by then able to recognise, examples of the Rules of the Funnel, thereby making things worse instead of better. Delegates frequently come up with several dozen such examples between them.

The experiment itself can be carried out with very simple apparatus:

1. A funnel, such as is to be found in any kitchen or garage;
2. A holder for the funnel, e.g. a desklamp to which the funnel can be fixed with wire;
3. A marble, small enough to pass through the stem of the funnel;
4. A table or other flat surface, horizontal and covered with a soft washable tablecloth, preferably ironed so as to have no creases;
5. A felt-tip pen with non-permanent ink; and
6. A ruler and protractor, or other devices for measuring distances and angles, not necessarily with great accuracy.

A target is marked on the tablecloth, and the funnel is positioned above the target. The marble is dropped through the funnel, and the position is marked on the tablecloth where the marble comes to rest. The holder, and hence the funnel, may then be moved according to one of a variety of Rules which we shall stipulate below. The marble is dropped through the funnel a second time, the position where it finishes is marked, and the funnel is moved again. The process should be repeated a few dozen times. What Rules for moving the funnel shall we consider? Deming suggests four.

Rule 1 is the easy and lazy one: don't shift the funnel, irrespective of where the marble comes to rest. Computer simulations of 100 drops of the marble under Rule 1 are shown in Figure 13. Not surprisingly, we see a roughly circular pattern centered on the target.

FIGURE 13
100 Marble-Drops Using Rule 1 of the Funnel

Well, that's not good enough. Let us try to improve matters:

> *"Let's do something. Don't just sit there.*
> *Do something about it.*
> *Move the funnel."*

Rules 2 and 3 move the funnel in attempts to compensate for the amount by which the marble misses the target. We shall describe these Rules in reverse order since Rule 3 (using Deming's numbering) is a relatively crude attempt at compensation, whereas Rule 2 is more refined. Rule 3 operates as follows. Suppose the marble finishes up six inches *east* of the target. Then the funnel is aimed six inches *west* of the target for the next drop. Or, if the marble finishes up four inches *south-west* of the target, the funnel is then moved to aim at the point four inches *north-east* of the target for the next drop.

The obvious weakness of Rule 3 is that it pays no attention, when specifying the next position of the funnel, to where the funnel was aimed for the current drop. The consequences of this are easy to see when carrying out the experiment. The reader might, however, care to picture the operation and try to deduce what the behaviour due to Rule 3 will be before looking at Figure 15.

Rule 2 adopts the more sensible policy of moving the funnel *relative to its last position* rather than relative to the target. So, returning to the above illustration, suppose the marble finishes up six inches east of the target. Then Rule 2 moves the funnel six inches west *of its current position*. And if the marble is four inches south-west of the target then the funnel is moved to four inches north-east *of its current position*.

Figures 14 and 15 show the patterns of 100 drops using Rules 2 and 3 respectively. Rule 3 indeed gives quite horrendous results. As time progresses, the general trend is for the marble to move further and further from the target, and to oscillate on consecutive drops from one side of the picture to the other:

> ### *"Rule 3. Oscillation, back and forth,*
> ### *eventually ever wider and wider*
> ### *to explosion."*

The reason for the oscillation is this: if the funnel is, say, three feet *east* of the target, the marble will presumably finish up somewhere around that same position—which implies that Rule 3 will then move the funnel to roughly three feet *west* of the target for the next drop. After that, it will be back over to the *east* again, and so on.

And so we look, with eager anticipation, to the results from the "refined" Rule 2. But what a disappointment! Agreed, things are nowhere near as bad as with Rule 3: indeed, we are back to a roughly circular pattern around the target. But the circle is *bigger* than it was—i.e. more variability, poorer quality. In fact (although this cannot be calculated exactly from the simulation), any reasonable theoretical way of measuring the areas of the two circles produces the result that the area of Rule 2's circle is *double* the area of Rule 1's circle.

So that great idea has failed. What to do? Perhaps we should forget all about the target itself and, in the interests of improving quality, concentrate instead on minimising the variability between successive drops. That way we can at least improve uniformity and consistency, though centered on some position other than the initial target. There is an obvious way to do this. (The reader will not be able to see the tongue firmly placed in my cheek as I write, so I had better warn you that that is where it is!) It gives us Rule 4 of the Funnel: at each stage, place the funnel directly over the position where the marble has just landed.

Well, one part of that description is valid: Rule 4 *does* minimise the likely distance between successive drops. So, in the very short term, this Rule does seem to have some merit. But beware! What about the long term? Figure 16 provides the answer. The behaviour is virtually as bad as that of Rule 3:

"The system explodes."

As time proceeds, the tendency is for the marble to get further and further away from the target. That is hardly surprising, seeing that the target doesn't figure in the calculation of where to place the funnel! The only real difference from Rule 3 is that the movement does not swing from one side of the picture to the other: it just tends to drift away in one general direction:

"All that work to make things worse!"

FIGURE 14
100 Marble-Drops Using Rule 2 of the Funnel

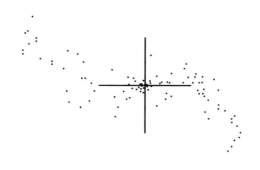

FIGURE 15
100 Marble-Drops Using Rule 3 of the Funnel

As regards practical illustrations of the Rules of the Funnel, the fact that the Ford compensation tool example is a case of Rule 2 should now be clear: if the diameter of the shaft is 0.10 mm too big, the process mean is adjusted down 0.10 mm (from its current value) in order to try to get the next one closer to nominal.

Any form of systematic item-by-item compensation is a candidate for Rule 2. Deming refers to this in the paragraph

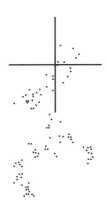

FIGURE 16
100 Marble-Drops Using Rule 4 of the Funnel

beginning at the bottom of page 141 in *Out of the Crisis*. Calibration procedures are often good examples. A standard item is measured at the beginning of each day, or of each shift, and the instrument is adjusted according to the observed error. On page 330 of *Out of the Crisis*, a delegate at a four-day seminar tells of an analogous example involving adjustment of sights on a submarine weapon according to the amount by which the first shot of the day misses its target.

Having established the idea of Rule 2, it is then not a big step to extend the same "tampering" concept to less precise but often more serious applications. Workers who are praised or blamed according to whether their measured performance happens to be above or below average, when in fact their work is in statistical control, are being inflicted with Rule 2 or possibly one of the other Rules, and overall performances will accordingly become more variable. In Chapter 18 we shall see how Deming implies that immediate direct action on faults, complaints, mistakes, accidents, etc. are also often cases of Rule 2 (or worse).

Some words of clarification are needed at this point. Of course, as pointed out in Chapter 4, it is true that reactions to accidents, errors, failures, etc. must sometimes be justified. Indeed, action to try to pacify or compensate the customer or others involved is likely to be necessary or, at least, advisable. The type of action on which Deming is casting doubt here is action on, or affecting, *the process or system* as a result of the occurrence in question.

So how does that fit into the theory of the Funnel Experiment? The answer is that there are two fundamental assumptions underlying the Funnel Experiment, assumptions which are not always met in practice. One assumption is that it is actually possible to aim the funnel at the target, or at any other specified objective; the second is that the process producing the errors from that objective is *in statistical control*. If the former assumption is not satisfied, the article by Frank Grubbs (referenced in the footnote on page 329 of *Out of the Crisis*) becomes appropriate. Deming remarks that Dr. Grubbs' work is not a solution for the Funnel. Indeed it isn't: it is a solution to the problem which arises when only the second of our two assumptions is true.

On the other hand, if the second assumption is false and the first is at least roughly true, Rule 2 can produce benefit compared with Rule 1; i.e. it *can* then be relatively helpful to adjust item-by-item. In particular, if the process average is wandering around unpredictably, Rule 2 in effect keeps chasing after it and, especially if the average doesn't move by enormous amounts between steps, the chase will not be entirely forlorn. Examples of such "wandering" special causes are catalytic decay and fluctuations in ore purity. Of course, the results will still be nowhere near as good as if the process could be brought into control and Rule 1 then applied.

The crux of the matter therefore is whether that underlying process is in control or not. If it is, then Rule 2 is harmful; if it is not, then Rule 2 could be helpful. So the important point to study as regards reacting to accidents, mistakes, complaints, etc. is whether or not they are *part of the system*—i.e. are they the results of exceptional special causes, or are they, in effect, relatively high or low points which are nevertheless between limits on a control chart? In the former case, action does need to be taken to identify the special causes and to try to prevent their recurrence; but, in the latter case, direct action on the basis of those occurrences is just like the action of the Ford compensation equipment.

Practical illustrations of Rule 3 are less easy to find. Two likely reasons are that, as we have already seen, Rule 3 is a particularly silly rule—being based on similar notions to Rule 2, but in ignorance and neglect of where the funnel was last aimed. Also, again as we have seen, the effects of Rule 3 are so berserk that, in practice, such behaviour will generally be noticed, even if not understood, and a different strategy adopted. But cases of Rule 3 are by no means impossible to discover. When firing a rifle, suppose the bullet ends up one inch above the target. It is not inconceivable that the decision is then taken to aim one inch below the target, *rather than one inch below the point being aimed at previously.* Stupid, yes, but not inconceivable.

A useful clue for recognising cases of Rule 3 can be the observed behaviour—i.e. increasing instability with wild swings to and fro. Because this behaviour is so obviously undesirable, it tends to be accepted in practice only if the time-intervals concerned are fairly large. One might think of swings between the political left and right or between libertarianism and censorship.

With the harm done by Rule 2 being limited (but by no

means negligible), and with Rule 3 being more rare, Rule 4 may well be the most serious in practice. Rule 4 is insidious. In a very localised sense, it does indeed reduce variability. If the two assumptions just discussed are true, one cannot argue against the fact that the way to minimise the average discrepancy between any one outcome and the next *is* to place the funnel over that first outcome. But the long-term effects of that localised optimality have already been seen in Figure 16.

An easy way to understand what goes on in Rule 4 is to think of the game of "Gossip" or "Chinese Whispers" sometimes played at parties. One person whispers a sentence or more to a second person, who then whispers his understanding of what he has heard to a third person, and so on. By the time the sentence has passed between 15 or 20 people, it is usually very different from how it started.

Deming mentions three examples of Rule 4 on pages 329–330 of *Out of the Crisis*. First, there is the operator who tries to achieve consistency by attempting to make every piece like the last one. Second, and closely related, is the practice of colour-matching, where each batch of material or paint or developed photographic film is matched as closely as possible to the previous batch. Third is the practice of "worker training worker." People on the job train a new recruit; in three day's time, he's a veteran and is deemed ready to train further newcomers, who themselves train more new blood a few days after that. Why is this done? One reason is that it is (apparently) cheap. With regard to the sixth of the 14 Points:

> *"Management work with nickels and dimes;*
> *they ignore the huge losses."*

More laudably in principle, though not in practice, it may be genuinely believed that someone on a job knows more than anybody else about that job. So he does, as regards some of it. But he also knows some that is wrong. The idea of worker training worker sounds great but, as time progresses, the result is to move further and further away from what is wanted. We are in fact involved, in Rule 4, with what statisticians term a "random walk"—a random walk away from the target. The practice of worker training worker could be even worse: training requires training, and without it there will be additional bias away from the target.

Still more horrific effects of Rule 4 occur at the higher levels of management:

> *"Still worse,*
> *executives working together,*
> *without guidance of Profound Knowledge.*
> *They go off by Rule 4 toward the Milky Way,*
> *only doing their best."*

Finally, let us briefly mention some other examples amongst the many both in *Out of the Crisis* and those raised in seminars by Deming or the delegates. It is not always clear which of the Rules is most appropriate. What does become clear is that they are all examples where, however sensible the action might seem in the short term, the long-term effect is to make things worse. Further examples from *Out of the Crisis* concern making workers responsible for the defectives they have apparently produced (pages 249–250), language teaching (pages 254–255), and consumption of water in a production process for refined sugar (pages 390–391). The third of the Obstacles in Chapter 3 was the "Search for Examples"; the harm caused by copying examples is discussed at some length in Chapter 16, and is an illustration of Rule 4. A familiar case nowadays is that of benchmarking.

Once the message sinks in, delegates at the four-day seminars come up with numerous examples of inappropriate action on effects rather than causes, including blame on innocent parties who have nothing to do with the causes—cursing at a waitress for bad food, and making a hotel maid responsible for missing towels, for examples. Other suggestions have concerned movements in interest rates, safety and security issues instigated by isolated occurrences, forecasting, budgets—including the need to spend up to budget or lose out the following year, salaries tied to the inflation index, revising a course as the result of some student feedback, quarterly readjustment of assets and manpower for quarterly reports, under-bidding competitive prices, scheduling the length of the next meeting according to the length of the current one, getting a consultant in to sort out the effects of a problem rather than really finding out why it occurred, stopping a production line as soon as something goes outside specifications, photocopying a photocopy of a photocopy of ... , adjusting the temperature on a thermostat, trade barriers, and nuclear proliferation.

The Funnel Experiment is a great example of how a very simple tool can teach deep and profound ideas. Do not let its simplicity fool you—the messages it communicates are crucial.

Chapter 6

THE EXPERIMENT ON RED BEADS

Dr. Deming has been using the basic equipment for the Experiment on Red Beads for many years; indeed, he was using it in his first lectures to the Japanese in 1950 to demonstrate differences between common and special causes of variation. That basic equipment is a container of white and red beads in proportions of about 4:1, and a rectangular piece of plastic, wood, metal, etc., usually called a *paddle*, into which are bevelled 50 depressions. A sample of 50 beads is taken by inserting the paddle into the container, and then withdrawing it. (Note to statisticians: I deliberately do not say a *random* sample, even if the beads can be well-mixed before the sample is taken.)

The basic form of the Experiment on Red Beads, as demonstrated in the four-day seminars, has stayed relatively unchanged for several years. Volunteers are invited from the audience:

6 willing workers—no education needed: training will be provided; must be willing to obey orders without question or complaint (after the training is complete);

2 junior inspectors—must be able to count to 20;

1 Chief Inspector—must be able to compare two numbers for equality or otherwise, and to speak in a loud, clear voice;

1 recorder—must be able to write neatly and carry out simple arithmetic. (Has to be smart, therefore preferably female!)

A day's work for each willing worker is represented by his selection, using the paddle, of 50 beads from the container. White beads are good product, acceptable to the customer; red beads are not. A goal is set of perhaps one, or perhaps three, red beads, according to the foreman's whim or a message from higher authority. The workers are trained by the foreman (Dr. Deming) and are given precise directions on how the work is to be carried out: how the beads are to be mixed, plus directions, distances, angles, and agitation levels for using the paddle. The procedure must be standardised and regimented in order to minimise variation. The workers must take great care to follow all instructions; continuation of their job depends upon their performance:

"Remember, any day may be your last,
depending on your performance.
I hope you enjoy your work!"

The inspection process is overmanned but mostly effective. Each worker carries his day's work to the first junior inspector, who counts in silence and writes down the number of red beads, and then to the second junior inspector, who does likewise. The Chief Inspector compares, still in silence, the two counts. If they differ, there may be a mistake! More thought-provoking is the fact that, even if they agree, there may be two mistakes. However, the operational procedure is that, if the counts differ, the inspectors must recount, still independently of each other. When the counts agree, the Chief Inspector announces the count, the recorder notes it on an overhead projector transparency, and the worker returns his beads to the container—his day's work is complete.

The work continues for four days—24 drawings in all. The foreman comments continually on the results. He congratulates Al for getting the number of red beads down to four, and the audience applauds. He castigates Audrey for reaching 16, and the audience laughs nervously. How could she possibly make four times as many defectives unless she is careless or idle? Nobody else should feel any satisfaction in their performance either: if Al can make four, *anybody* can make four. Al is clearly "worker of the day"; he will receive a bonus. But, the next day, Al makes nine red beads: he's getting complacent. Audrey gets ten—she was off to a bad start, but now she's really trying to improve, especially following the tough talking-to she received from the foreman at the end of the first day. Hold it—stop the line! Ben has just made 17 red beads. Let's hold a meeting and have a discussion to try to find out what went wrong. That kind of performance will put us out of business. At the end of the second day, the foreman has a serious talk with the workers. As people become familiar and experienced with the job, performance should improve. Instead, following a total of 54 red beads on the first day, 65 were produced during the second day. Don't the workers understand the job? The job is to make white beads, not red ones. The future looks bleak. Nobody has reached the goal. They must try harder.

The workers, despondent, chastened, return to the task. There are a couple of bright spots. Audrey improves further, getting down to seven red beads. And Ben gets back on track, repeating the nine red beads that he had on the first day. But everybody else gets worse. The total of reds goes up further, to 67. The end is nigh. The foreman tells his workers that if there isn't significant improvement then the place will have to close.

The fourth day commences. With relief, we find that things get better—led by Audrey who now produces only six red beads.[1] But the day's total only gets down to 58 red beads, still worse than the first day of all.

Here are all the results so far:

	Day 1	Day 2	Day 3	Day 4	**Totals**
Audrey	16	10	7	6	**39**
John	9	11	12	10	**42**
Al	4	9	13	11	**37**
Carol	7	11	14	11	**43**
Ben	9	17	9	13	**48**
Ed	9	7	12	7	**35**
Daily Totals	**54**	**65**	**67**	**58**	
Grand Total:					**244**

At this stage, the foreman may decide to invoke a great contribution to management—to keep the place open with the best workers. He fires Ben, Carol, and John, the three workers who have made 40 or more red beads over the four days, and he retains Audrey, Al, and Ed, pays them a bonus, and puts them on double-shift.

Not surprisingly, that doesn't work either.

We have a rare privilege in the Experiment on Red Beads —we understand the system well enough to be pretty sure that it is in statistical control. Once we realise that, we confirm the pointlessness of all the reactions of the foreman (and everybody else) to the results supposedly produced by the workers but, in

---

[1] Note to traditional statisticians: under the standard null hypothesis, and given that Audrey gets four different scores, the chance that she improves every day is $1/4! = 1/24 = 0.042$. That's a significant result at better than the 5% level of significance!

actuality, entirely due to the system itself. They have been reactions to mere random variation.

Suppose we didn't have that understanding of the system. What would we do? We would plot the data on a control chart, and let it tell us about the behaviour of the process. The Central Line on the chart is placed at the average count, i.e., $244 \div 24 = 10.2$. Then the calculation for σ is:

$$\sqrt{10.2 \times \left[1 - \frac{10.2}{50}\right]} = 2.8;$$

so the Upper and Lower Control Limits are placed at:

$$10.2 + (3 \times 2.8) = 18.6 \quad \text{and} \quad 10.2 - (3 \times 2.8) = 1.8$$

respectively. (See similar computations on page 347 of *Out of the Crisis*.) The control chart is shown in Figure 17.

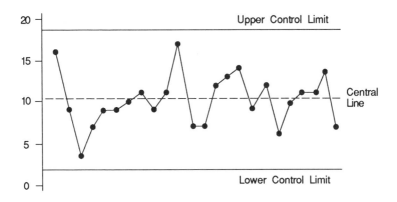

FIGURE 17
Control Chart of Data From the Experiment on Red Beads

This chart bears out what we believed—the process is in statistical control. The variation is caused by the system. The workers are helpless: they can only provide what the system will provide. The system is stable, predictable. If we run the experiment again tomorrow, or the next day, or next week, we are likely to obtain a similar spread of results.

To delegates actively looking for messages from the Experiment on Red Beads, there is much to see, even before Deming commences his summing-up. They see pleasure at good results and embarrassment at bad ones, hardly helped by the foreman's criticisms and threats. They see trends (like Audrey's statistically significant improvement), they see relatively consistent performances (like John's) and erratic ones (like Ben's); they see and hear complaints from the foreman when worthless and meaningless directions are not carried out to the letter. They see workers compared with each other when they are in fact helpless to affect their results—results which are entirely governed by the system within which they labour. They even see workers losing their jobs, through no fault of their own, while others are rewarded for no special virtue other than that the system has treated them with generosity.

Deming points out some of these features of the experiment, plus a few more subtle ones. The cumulative averages at the ends of each of the four days are respectively :

$$54 \div 6 = 9.0;$$
$$119 \div 12 = 9.9;$$
$$186 \div 18 = 10.3; \text{ and}$$
$$244 \div 24 = 10.2.$$

He asks the audience what they think this average would settle down to if the experiment were carried on at greater length. Remembering the 4:1 ratio of white to red beads, those knowledgeable of the mathematical law of large numbers, amongst others, are clear that the answer must be 10.0. Not so. It would be so if the sampling were done with random numbers. But, instead, it is done by dipping the paddle into the container. That is *mechanical* sampling, not *random* sampling for which the mathematical law *would* apply. If evidence is needed, Deming quotes figures obtained from using four different paddles over the years. Two of these at least are results which the traditional statistician would judge as "significantly different" from 10.0. And what kind of sampling does one carry out in industrial processes? Mechanical or random? Where does that leave those who depend only on standard statistical theory in industrial applications?

Not everything about the experiment is bad. The inspection process has an important good aspect about it. At first sight, it seems to fall foul of an issue that Deming sometimes raises at his seminars: the problem of *divided responsibility*. But the junior inspectors' contributions here are kept strictly independent of each other; the hazard of divided responsibility is the hazard of *consensus*. The matter is discussed at greater length in Chapter 21. (Consider also Rule 4 of the Funnel.)

The question naturally arises, both in the Funnel Experiment (Chapter 5) and in the Experiment on Red Beads, as to what *can* be done to improve matters? Just because Rule 1 is better than Rules 2, 3, and 4, does that mean we just sit back and do nothing? Or course not! We know the answer already. Since the underlying system is in statistical control, genuine improvement can only be produced by genuine change to the system itself. As Deming puts it:

"Improve Rule 1!"

Improvement cannot be produced by tampering with output, i.e. managing by results (MBR): for action based on results can only be appropriate in the presence of special causes. That is what Rules

2, 3, and 4 of the Funnel attempt, and that is what is attempted by all the foreman's interjections in this Experiment.

Action on the system (tackling common causes) is usually more *difficult* than action which could be justified if special causes were present. In the Funnel Experiment, the funnel might be lowered, or a softer tablecloth might be used, damping out some of the marble's motion after it is dropped. In the Experiment on Red Beads, somehow the proportion of red beads in the container must be reduced—by improvements further upstream in the process, or in supplies of raw materials, or both.

Deming refers to his Experiment on Red Beads as "stupidly simple." So it is. As with the Funnel Experiment, the messages learned are not.

Chapter 7

OPERATIONAL DEFINITIONS

There is a whole chapter (Chapter 9) on operational definitions in *Out of the Crisis*. However, somehow the topic does not generally receive the attention which Deming clearly believes it should. As early as the second sentence of Chapter 9, he writes: "No requirement of industry is so much neglected" as the use of operational definitions. He also points out that, whereas the topic is studied in courses on philosophy and the theory of knowledge, it is not often covered where it is most needed—in courses on business, engineering, and the natural and physical sciences.

There is no doubt that Deming considers work on operational definitions to be of supreme importance. He gives two particular historical pointers in support of this assertion. First is the apparently stunning fact that Shewhart believed his work on operational definitions to have been of greater importance than his development of the theory of variation and of the control chart. Second, Deming knows that the Japanese paid great attention to the development of operational definitions in the early 1950s, and realises that the benefits they gained from this bore comparison with those obtained by their use of the concepts and tools of the statistical control of processes. But is this so "stunning," after all? Remember the words at the beginning of Chapter 4: the "message for management" is all "to do with reducing variation." It should soon become apparent that the use of operational definitions has a great deal to do with reducing variation.

What is an operational definition? Combining the words of Shewhart (in his 1939 book) and Deming, it is a definition *which reasonable men can agree on and do business with*. All meaning begins with concepts—thoughts, notions, images in the mind. Some dictionaries emphasise that a concept is abstract or theoretical. Deming's word to describe a concept is "ineffable," i.e. unspeakable, beyond description. One can hardly do business with concepts!

Unfortunately, many try, without either realising that that is what they are doing or understanding the hazards. How many people rush into print without a thought of whether what they are saying will be comprehensible to others? How many reports, or instructions, or procedures get written down which are comprehensible only to the writer? (Manuals on computing seem especially prone to this failing!) Why should the writer take greater care? After all, it is perfectly obvious to *him*. He should take care because presumably the act of writing means he wishes to communicate with others. Some, at least, of those others may not be on his wavelength. How many arguments, in and out of court, have taken place because contracts contain concepts rather than operational definitions: concepts which "are open to inter-pretation?" The very nature of an operational definition is that it is *not* "open to interpretation"—instead, it interprets:

> *"Words have no meaning*
> *unless they are translated into action,*
> *agreed upon by everyone.*
> *An operational definition puts*
> *communicable meaning into a concept."*

On the surface, maybe the idea of operational defini-tions does not sound too demanding. But try it. As soon as we attempt to turn ideas into practicalities (which is surely what operational definitions are all about), the trouble starts:

"Practice is more exacting than pure science; more exacting than teaching."

Ambiguities caused by lack of operational definitions are legion. For example, with reference to Chapter 1, what is an export? Its official meaning is certainly not what Deming thought it was! Concerning figures for fires in a business establishment (*Out of the Crisis*, page 324), what constitutes a fire? What constitutes improvement in a patient's condition (*Out of the Crisis*, page 443)? What is a typing error (*Out of the Crisis*, page 226)? What is a wrinkle (*Out of the Crisis*, page 290)? What is meant by "on time" (*Out of the Crisis*, page 475)? Deming often questions his students: What is a "bestseller"? What is "clean"? We shall need to know if our job is to wash the table. Does it mean clean enough to eat off, or to sell, or to operate on? What is "satisfactory"? Satisfactory for what? Satisfactory to whom? What test should we apply? In Chapter 24, we shall see the need to know what is meant by " ... careful, correct, attached, tested, level, secure, complete, uniform ... " and so on. How can we do our job if we don't know? In Chapter 29, it will appear that the definitions of sales and accidents are highly dependent on management pressures. One of Deming's most famous examples is to ask: "What means a label on a blanket reading '50% wool'?" For answers and discussion, see *Out of the Crisis*, pages 287–289!

My friend Malcolm Gall of Hydro Polymers reminds me that in 1986 one politician claimed that unemployment in Britain stood at three million, while another placed it at four and a quarter million. They were both right—according to their different definitions of what is meant by "unemployed." Somebody else told me that the official definition of "unemployed" has been changed over 30 times since our Conservative Government came into power in 1979.

I am also indebted to Malcolm Gall for the following two examples. We hear of the need for "Zero Defects." What is a defect? For example, what is a surface with "no cracks"? What is a crack? Cracks too small to be seen by the naked eye may not be counted. But *whose* naked eye? Do all naked eyes have the same power of definition? And are all marks, depressions, irregularities to be counted as cracks? Give the observer a magnifying glass, and further cracks may be identified. Give him a microscope, and even smaller cracks can be seen and counted and reported as defects.

What is meant by material containing "no cadmium"? What is "no"? The analytical technique to be used will have its limit of what it can detect. To an analytical chemist, "zero" does not have a practical meaning. The noise in the electronic measuring devices and the losses in preparation of samples will often make it impossible to distinguish between different quantities of impurities. In some countries and international corporations, "cadmium-free" material has been defined as that containing no more than 75 parts per million by weight. Further, an unambiguous method of measurement is needed to decide whether or not a quantity of material does or does not satisfy the criterion. Is it sufficient to test a sample? How is the sample to be selected?

The opportunity to discuss virtually anything in depth uncovers ambiguities that even the experts did not know existed.

Do we need specifications? If so, we surely need operational definitions of them:

> *"It's pretty important that we understand these*
> *things if we're in business.*
> *If we're not in business,*
> *then it doesn't make so much difference."*

What is "punctual"? An eye-catching headline from the *Daily Mail*, London, 16 July 1987, tells us that "British Rail

Redefines Punctuality." The report reads as follows:

> "British Rail is redefining punctuality, to make
> it appear that more of its trains are on time. ... At
> present, trains that arrive within five minutes of
> their scheduled time are said by BR to be punc-
> tual. In future, trains that arrive within 10 min-
> utes of their proper time on Inter-City services
> will be listed as 'punctual'."[1]

Again, a *method* of measurement is also needed—how precisely
do we record the time of arrival?

I showed this press-cutting to Dr. Deming. He smiled.
"Why ten minutes?" he asked. "Why not 15? Why not 30? Why
not more? Sounds a great way to get Zero Defects!"

Perhaps the moment when the newcomer suddenly realises there really is much more to all this than meets the eye is
when Deming suddenly comes out with:

"There is no true value of anything."

(A qualified version of this statement appears as Item 9 in Part C
of the System of Profound Knowledge in Chapter 18.) Moreover:

---

[1] A stark comparison between this somewhat cavalier attitude to punctuality on the railways with the situation in Japan is seen in the two stories which Deming tells on page 27 of *Out of the Crisis*. I am grateful to
Brian Read, the BDA's Director of Advisory Services, for spotting a similar example in a letter from a Mr. Robin Leale, published in *The Times*,
London, 1 January 1990: " ... On a Saturday outing to the shrines at Nikko
involving an hour or so's train journey from Tokyo, a member of our
party was concerned about alighting at the correct station. The advice
from our thoroughly conscientious tour-guide was: 'Please do not worry
yourself about the name of the station. Just get off the train at 10:46
am!'" As a final thrust, Mr. Leale wrote: "Of course, to be quite sure of a
safe arrival, one also needed to have an equally reliable Japanese watch
to hand." The reader is again reminded of the Deming quote which
commences our Chapter 4.

> *"A physics book defines experimental error*
> *as the difference between*
> *the observed value and the true value.*
> *Wrong—such ignorance!"*

Let's think about "no true value" with the aid of a simple example. Animal crackers! The question is: how many of the animal-shaped biscuits are there in a particular packet? The traditional statistician would not give this question a second thought—that would be "trivial." He would immediately start considering the way that the number *varied* from packet to packet, thinking about means, standard deviations, and distributions. A somewhat better statistician would first think about stability. But, in truth, deciding on the number in any given packet is not an easy task. For biscuits get broken. Some may be halved, others may have corners broken off. Is a cow with three legs to be counted as a cow? How about one with no legs? How about a leg? Of one thing we can be sure—every piece of biscuit in that packet will be different from every other piece. How do we define what counts?

If there is no true value, what is there? *There is a number that we get by carrying out a procedure*—a procedure which needs to be operationally defined. But *if we replace the procedure by another procedure*, also operationally defined, *we are likely to get a different number.* Neither is right, neither is wrong. If the procedure is not operationally defined, we are likely to get different numbers even with the same procedure.

Suppose we have an operational definition. Is it "right" or "wrong"? No. The question is: does it do what we want it to do?

What is the number of people in this room? What is the number of people in this building? Do we include those who have just gone out to lunch, or who leave early, or who are just on the point of entry into the building? Do we include the newborn baby

114

the proud mother has just brought in to show her friends? Do we include the deliveryman just bringing in a parcel?

And if there is difficulty with that, how do we count the number of people in this city? Impossible? But we have to produce figures in the Census. Do we count students, either those studying here or those studying away but whose parental home is here? What about those without homes? What about those with two homes? What about refugees? What about those in the doss-houses?[2] They live here, if we can call it living. But will they be included in the Census? It's pretty expensive. Deming quotes $100 as the cost of each such inclusion; and $200 for those sleeping on park benches, in the subways, etc..

An early example which Deming shows to illustrate his arguments concerning operational definitions and "no true value" uses figures, dated 22 December 1955, on the percentage iron content of iron ore mined by the Yawata Steel Company. The table he shows is as follows:

Ore	Class	Old Method	New Method
Dungun	A	59.95	55.33
Larap	B	56.60	55.30
Larap	C	59.25	58.06
Samar	D	55.55	50.42

In the old method, the ore was chosen by scooping samples off the top of the trucks; in the new method, the samples were taken off the conveyor belt. Neither method is wrong, neither method is right; neither set of figures is wrong, neither set is right. The question is: does either serve the purpose better? If so, use it.

Deming also illustrates these matters by referring to the history of determinations of the speed of light, the data having

---

[2] American translation: shelters.

been collected and published by Shewhart in his 1939 book. It will be no surprise to learn that the values are all different. The reader should refer to *Out of the Crisis*, pages 280–282 for the data and Deming's discussion:

> "It comes as astonishment to most people that there is no true value for the speed of light."

He also comments on a much earlier observation of the speed of light (Galileo in 1606) to the effect that:

> "If the speed of light is not infinite, then it's awful damn fast!"

Incidentally, recall our observation in Chapter 6 that the long-term average (even if *that* could be defined) number of red beads obtained in the famous experiment is not directly related to the proportion of red beads in the container. That comes as quite a shock to most statisticians and others. But how *could* it be so related if (as *is* the case) hundreds of usages of two different paddles produced averages of 9.4 and 11.3 respectively (see *Out of the Crisis*, pages 351–352)?

Whereas there is no true value in practice, there can be true values in mathematical theory. Many readers will know that, in mathematical theory, one can prove that the ratio of the circumference of a perfect circle to its diameter is π, an irrational constant whose first few figures are 3.14159265. I believe that modern computing power has enabled its calculation to some thousands of decimal places.

However, let us be clear on the conditions under which this value is "true." First, we need a method of measurement which is of infinite accuracy, able to measure both straight lines and curves. Second, we need a perfect circle. Third, both the line representing the diameter and the curve representing the circumference of the circle must have zero thickness. Under these conditions, π is the true value of the ratio.

But that is the world of mathematical theory. Perfect circles are ineffable, perfect measurements are ineffable, zero thickness is ineffable. In practice, of course, not one of these conditions is met, nor could ever be. However we define and measure the ratio: circumference ÷ diameter, we cannot and never will be able to obtain that value of π. Maybe we can measure to three places of decimals; say, the circumference is 6.237 cm and the diameter is 1.985 cm. Dividing the one by the other gives 3.142065... . Is that right or wrong? It's not equal to the "true value" of π. Does that make it wrong? If so, we are destined to always obtain wrong answers for the rest of our lives. Is it right? Then "right" does not mean "equal to the true value." No; to repeat: there is a number that we get by carrying out a procedure. But if we change the procedure, we are likely to get a different number. So what is "true"?

Finally, I must confess I used to be perplexed by Deming's reference to a point lying outside 3σ control limits as constituting an "operational definition of a special cause" (see Chapter 4). Showing my mathematical upbringing, I used to think simply of common causes as coming from within the system, and special causes as coming from outside. I still mention this idea in introductory presentations, as I have done in Chapters 2 and 4, for it is a useful *concept*. But, in practice, the situation is nothing like as clear-cut. Maybe a few causes of variation are "obviously" common and others are "obviously" special; the truth is that there are usually many more whose role is arguable. Also recall that, in terms of that conceptual definition, there exists *no* real process unaffected by special causes (*Out of the Crisis*, page 334), i.e. no real process is in statistical control. Even if that were not the case, how would we know? So, in *concept*, that definition of common and special causes may be fine; in *practice*, it gets us nowhere. We therefore need an operational definition for a special cause or, if you like, an operational definition of *when to look for* a special cause:

"Shewhart contrived and published the rule in 1924
—65 years ago.
Nobody has done a better job since."

Chapter 8

PROCESSES AND SYSTEMS

We have already referred many times in this book to *processes* and *systems*. These are words which trip lightly enough off the tongue, but it will be wise at this stage to take stock of what we really mean by them.

What is a process? I have heard it defined as anything which can be described by the use of a present participle, preferably with an object! The present participle is the form of a verb ending in *-ing*, e.g. "writing." To provide an object obviously helps to clarify what is being considered, e.g. writing this book, writing a report, writing a letter, writing a word.

The fifth of the 14 Points speaks of improving "constantly and forever every process." Our definition in the previous paragraph doesn't help us much with that! A vitally important step which does help is to become very conscious of the fact that every process has inputs and outputs (or outcomes)—see Figure 18. The process executes some kind of change; some of the inputs contribute to the activity of change, and some of them get changed.

Almost always, the inputs to a process are far more numerous and varied than is normally realised, and often the same is true of outputs. This is well-illustrated by Figure 19. This Figure is based on a schematic which is used repeatedly in Bill

Scherkenbach's *The Deming Route to Quality and Productivity* and elsewhere.[1] This is not to say that *every* process has to have every one of the six types of inputs and outputs shown in that schematic. However, in the case of inputs, it is often easy to identify examples of all of the six types; the issue is much bigger and broader than we usually take account of.

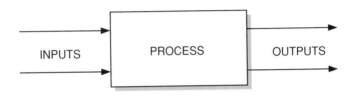

FIGURE 18
Basic Notion of a Process

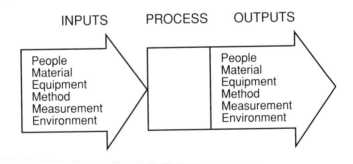

FIGURE 19
Basic Notion of a Process,
Showing Multiple Types of Inputs and Outputs
(Adapted from William W. Scherkenbach)

---

[1] I have added Measurement to the lists of inputs and outputs at the suggestion of Colin Nichols.

Have you ever used an Ishikawa fishbone (cause-and-effect) diagram? There is no need to go into any formal analysis here; many other publications, such as those providing training in the "seven simple tools of quality control," specialise in this kind of very useful decomposition of a process. Anyone who has participated in a fishboning exercise cannot have failed to be impressed by the number of cause-factors for a single effect which are generated in even a short brainstorming session. All of these cause-factors are suppliers to the effect, which is therefore their customer. A strong message to management is that, seeing there are usually so many, they should consider very carefully the justification or otherwise for jumping in at one of them as being the "obvious" single or main cause of the effect. Nowhere can that message be more pertinent than in the framework of performance appraisal (see Chapter 30).

Consider a selling process, for example—it might be a sale in a store, such as of a television set or some furniture, or one made through a visiting salesman, such as of some insurance or of a copy machine—or indeed of a whole batch of copy machines. Amongst the inputs to a selling process are, of course, the people most directly involved: the salesman and the customer (the one making the choice and signing the order form). But there are many more: the features of the equipment, the reputation of the brand, its price, the warranty terms, the advertising, the reputation of the distributing company, the mechanism for following up leads, the company's sales methods, the way the salesman is paid, the customer-company's purchasing policy, and so on. All these and many other influences affect whether and how the sale takes place. They are all *inputs* to the process, provided by a variety of *suppliers* to the process.

Notice the huge extent to which *management* supplies the process, i.e. supplies these and many other types of inputs to it. For example, the extent to which the salesman tries to push the customer into ordering something he doesn't really need is going to be strongly influenced by the levels of commission, bonuses, incentives, etc. which define the salesman's compensation. The whole business philosophy of the selling company affects the balance between the desires to satisfy or screw the customer.

How about *outputs* from the process and recipients of those outputs, i.e. the process's *customers*? Whereas the inputs and suppliers include anything and anybody which has influence on the process, outputs and customers include anything and anybody that is influenced by it. So the list clearly includes those people who will watch the television or will use the copy machines, etc.. But, more immediately, outputs from the process will almost certainly include a lot of paperwork, and customers (in the sense just defined) will include both companies' Administration Departments, and their own or related processes such as order-processing, delivery, accounting, finance, service, and supplies (e.g. paper and toner for the copy machines). The reliability of the equipment can certainly affect the local environment—think what traumas are caused, either in the office when the copy machine breaks down, or at home when the television fails!

Current styles of management tend to mitigate against the broad perspective of process-thinking: and this underlines the importance of the ninth of the 14 Points: Break Down Barriers between Departments. An extremely useful example in training sessions therefore (and, the more senior are the trainees, the more valuable the example becomes) is to get people to specify a process with which they are intimately involved, and then to start them thinking more comprehensively about its inputs and outputs, and hence about the suppliers and customers of that process. If a session is being attended by people from a number of different disciplines within the company, others might soon be adding to the lists of inputs and outputs, and this in itself can quickly lead to much better understanding of the process which is being studied.

So far in this chapter we have spoken only of *processes*.

Then what do we mean by a *system*?[2] Deming gives a very specific answer to that question:

> *"What is a system?*
> *It is a network of functions or activities*
> *(sub-processes or stages) within an organisation*
> *that work together for the aim of the organisation."* [3]

That is, a system can be thought of as a structure of interlinking processes. But many of what we refer to as *processes* can themselves be broken down into smaller sub-processes. So, surely, very small operations are referred to as processes, and very big ones as systems, but it is largely a matter of taste as to the scale of operation where one moves from the use of one word to the other. Related to the expanded vision of the inputs and outputs of processes as outlined above, there is bound to be a more substantial structure of interlinking sub-processes within processes, or of processes within systems, than one might initially think. I like the

---

[2] Very many views have been expressed on useful distinctions which can be made between the words *process* and *system*. For example, my friend Eve Williamson, Managing Director of Cambridge LINK, has correctly pointed out to me that many writers differentiate between the *system* as being what exists and the *process* as being how it operates. Pat Anderson, while he was Quality Assurance Manager in ICI Chemicals and Polymers, told me that he and his colleagues referred to the *process* as being the transformation of inputs into outputs, while the *system* was regarded as the complex of controls (particularly human and electronic) which supervise the working of the process. It does not seem to me that Dr. Deming draws such distinctions, and neither shall we in this chapter. On the other hand, we should recall that Deming often, though not always, uses the word *system* to signify a process which is in statistical control as opposed to one which isn't. Alternatively, he sometimes identifies the *system* with common cause variation, special causes then being "outside the system"—see, for example, the fifth of "Some Attributes of a Leader" (Figure 50 in Chapter 25).

[3] A smaller-scale example given by Dr. Deming is that "the mechanical and electric parts that work together to make an automobile or a vacuum-cleaner form a system."

picture shown in Figure 20, which is from Brian Joiner and his colleagues, as an illustration of this. Figure 20 is an example of a *flow-chart* or *flow-diagram*. Flow-charts are invaluable aids to

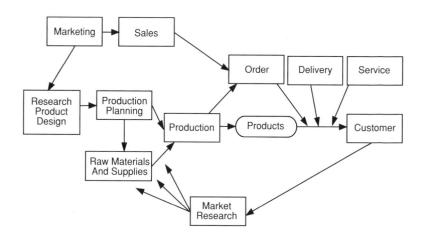

FIGURE 20
Flow-Chart of an Organisation

(Joiner Associates)

understanding processes—as regards both what is intended to happen, and what actually happens! I am amazed at the number of books, ostensibly covering tools for quality improvement, which hardly mention flow-charting, let alone develop the topic to any great extent. And some of those that do deal with it tend to concentrate far too much on some cosmetics, rather than on the real purpose and value of the technique. How can one hope to sensibly attempt improvement or innovation on processes without having a good understanding of what is happening now, in particular how the various components of the process interlink and interact with each other?—

> *"A flow-diagram is helpful*
> *toward understanding a system.*
> *By understanding a system,*
> *one may be able to trace the consequences*
> *of a proposed change."*

There are several different types of flow-chart in use, and even more may well be developed. A good introduction to some of them can be found in *The Team Handbook.* Benefits from flow-charting may come more quickly than expected: it is often the case that the very exercise of constructing a valid flow-chart for a process ("valid" implying what actually happens rather than what is supposed to happen!) leads straightaway to some considerable improvements.

Initially, people may well have different ideas on what the process consists of. Or they may have found that the official documented procedures are inadequate for the job and so, either out of necessity or at least the desire to do a better job, they alter the procedure. This can be another case, one of many in this book, where, as Deming puts it, "we are being ruined by best efforts." Everybody doing his best, with all the good intentions in the world, may well be counter-productive because there is absolutely no guarantee that local suboptimisation is going to be beneficial to the process as a whole.

Attempts must be made to develop knowledge of the ramifications on *all* the outputs from the process, not just the most obvious ones. Brian Joiner, for example, tells some stories about the way incentives to boost sales have been highly effective at doing just that, but have spelt disaster because the resulting demands have been beyond the capabilities of the administration, production, and delivery processes, etc. to support, and have also caused chaos to future planning (see Chapter 29).

As pointed out to me by Pat Anderson (already mentioned in a footnote earlier in this chapter), systems are unlikely to be well-defined in practice unless they are both suitable and adequate for the jobs for which they are intended, *and are written down in a way comprehensible to all involved.*[4] We see immediate links with our work on operational definitions (Chapter 7). One or more forms of flow-chart may be helpful, or the definition may be purely textual—but the system does need to be documented in some way. Yet again, there can be big differences between what is written down—the way the system is intended, or thought, to operate—and what actually happens. Incidentally, the acquisition of this latter information can be difficult if not impossible in an unfriendly work environment, i.e. one which is not being increasingly influenced by the 14 Points and other aspects of the Deming philosophy: Point 8 (Drive Out Fear) obviously comes to mind. But, even in a constructive environment, it may still turn out to be almost impossibly difficult to complete a flow-chart or other description because what is going on is so *un*systematic. To be blunt, if a system cannot be written down, it probably doesn't exist: that is to say it probably functions more on the basis of whim and "gut-feel" rather than on any definable procedure. This surely implies that the variation being generated is some scale of magnitude higher than really necessary, with the resulting (by now) well-known effect on quality. In general, even if it can be flow-charted, the greater the complexity of the system, the greater are the indications of trouble, and the greater is the potential for improvement and innovation.

Tim Fuller's work, particularly his article "Eliminating Complexity from Work: Improving Productivity by Enhancing Quality" has become celebrated in this regard. His best-known

---

[4] However, note that Deming does not consider it necessary to carry this advice to extremes: "Sub-processes need not be clearly defined and documented: people may merely do what needs to be done." In other words, we should keep a sense of proportion between documentation as a real aid and documentation for documentation's sake.

example is the comparison of a simple, but perfect, assembly system (Figure 21), i.e. where the kits can be relied on to be what they should be, with the way the system needs to be adapted if the kits of parts may be incomplete (Figure 22).

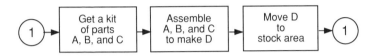

FIGURE 21
A System With No Errors, i.e. Kits Always Complete

(Timothy Fuller, "Eliminating Complexity from Work: Improving
Productivity by Enhancing Quality")

We would be well-advised to take note here that Deming's use of the word "system" is sometimes much more expansive. Indeed, rather analogously to Lloyd Nelson's concentration on the unknown and unknowable figures being the most important (*Out of the Crisis*, page 20), Deming's notion of a "system" may get so big that it becomes mind-boggling to even think of identifying all the most important inputs, let alone imagining having much influence (at least for some considerable time) on many of those inputs. The lesson to learn is to beware of suboptimisation—it is all too easy, and can be all too fatal, to plunge into suboptimisation of part of the process or system which looks relatively easy to handle, and forget about some of the consequences of that action, which may well be far more important than those which we do take into account.

The reader is referred to the section in *Out of the Crisis*, pages 317–318, headed "What is the system?". Deming gives a partial list of what might constitute "the system" as regards people in management—and, although the list is quite short, it contains substantial national and international components. More thought-provoking still is his reference to a comment by a delegate at one of his seminars pointing out that, to the production-worker, "the system is all but him."

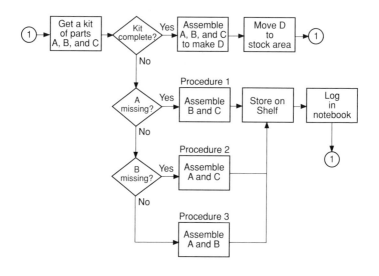

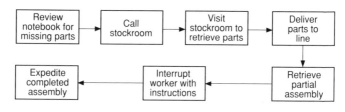

FIGURE 22

System Complexity Caused by Errors

(Timothy Fuller, "Eliminating Complexity from Work: Improving
Productivity by Enhancing Quality")

Such thinking also has its consequences on what is meant by processes and systems being in or out of statistical control. On the one hand, a small part of a system may appear quite stable, whereas the system as a whole is unstable; on the other hand, an out-of-control indication in a small part of the system may be well within the limits of variation of the larger system; in other words, what is special cause in terms of the part-system could well be regarded as common cause in terms of the larger system. This is yet further reason to think more broadly about interpreting the information which comes from control charts, and this is especially so in the case of human processes. It also shows even more clearly how absurd it is to think of describing either the mechanics or the theory of control charts in terms of probabilities (see again *Out of the Crisis*, pages 334–335). In addition, this bears direct relevance to Deming's belief that a main problem of the "Faulty Practices" of management (see Chapter 29) is that they result in substantial *tampering with stable systems*.

However, while remaining aware that the "system" can be interpreted over such a wide range of scale, we need to focus more closely on its interpretation in the context of our own organisation. Let us return to Deming's above definition of a system as a series of functions etc. *that work together for the aim of the organisation.* " ... work together ... " is key. The component sub-processes cannot, by themselves, accomplish the aim of the system. There is interdependence between them; they interact with each other; they cannot be separated:

"Management of a system therefore requires knowledge of the inter-relationships between all the components within the system and of everybody that works in it."

The greater the interdependence between the components, the greater is the need for communication and cooperation between them. Any attempt to improve the system without taking this fully into account is, at best, suboptimisation. It is, unfortunately, often the case that suboptimisation, i.e. optimisation of a sub-process, is incompatible with optimisation of the whole system (see "Example: Huge Financial Advantages of Cooperation" at the end of Chapter 15):

> *"It would be poor management, for example,*
> *to purchase materials at lowest price,*
> *or to maximise sales,*
> *or to minimise cost of manufacture*
> *or design of product, or of service,*
> *or to minimise cost of incoming supplies,*
> *to the exclusion of the effect on other stages*
> *of production and sales.*
> *All these would be suboptimisation, causing loss.*
> *All these activities should be coordinated*
> *to optimise the whole system."*

The incompatability between suboptimisation and optimisation has particularly unfortunate consequences because, since suboptimisation is a smaller problem and thus easier to understand, it is that which often underlies the judgments, decisions, and strategies of management. But:

> *"The performance of any component sub-process*
> *is to be evaluated in terms of its contribution*
> *to the aim of the system,*
> *not for its individual production or profit,*
> *nor for any other competitive measure."*

The degree of interdependence between component-processes varies considerably according to the type of system. In Chapter 3 we saw examples, such as music:

"A good example of a system, well optimised, is a good orchestra."

and football, in which teamwork is of the essence: this is because the degree of interdependence is so high. In other sports, such as ten-pin bowling or darts, "teams" are not required to work together in the same way: here it is more the case of scores from individual efforts being added together. But, even here, "team feeling" can still be a boost to individual morale and confidence, thus contributing to overall success. Again, the greater the interdependence between sub-processes, the greater the need for communication and cooperation (see Chapter 15). The degree of interdependence in a company is bound to be extremely high (whether this is recognised or not); the obvious conclusion follows.

So management's job is surely optimisation of the system —the system over which they have responsibility. But, in the same way that suboptimisation (optimisation of sub-processes) is often incompatible with optimisation of the system, optimisation of a system may well be incompatible with optimisation of a larger system which contains it. A further responsibility of management is therefore to look for opportunities to widen the boundary of their system for the purpose of better service and, yes, profit—NB: *both* of these, not just the latter; i.e. Win-Win, not Win-Lose (again see Chapter 15). It is in this context that Deming speaks of the desirability of monopolies and "fortresses of power" in Chapter 14.

As regards expansion of the boundary of a system, Deming writes as follows:[5]

---

[5] This is part of a communication from Dr. Deming, dated 3 January 1990.

"If the aim, size, or boundary of the organisation changes, then the function of the sub-components will, for optimisation of the new system, change. Thus, suppose that a railway company were to expand its service by addition of a motor-freight business (which could be done by forming a totally new business in motor-freight, or by buying an existing company that is already in the motor-freight business). The new system would be railway plus motor-freight.

The expanded system opens up new opportunities for service and profit. Optimisation of the expanded system would be different from separate optimisations. It might be wise, from the standpoint of service and profit for the expanded business, to optimise what is called piggy-back hauling—two trailers on a flat car, to be unloaded at destination and rolled away.

Optimisation of the expanded system would require knowledge of costs. Costs would vary with length of haul by truck at both ends, and length of haul by rail: also prediction of type and amount of business. Theory, possibly aided by simulation, would yield useful predictions for approximate optimum spacing between stations on the railway, and what hauls would be best made all the way by truck, rail not involved."

As we have seen, two particularly vital constituents of Deming's teachings to the Japanese from 1950 were developments in their understanding of (1) the nature of variation, and (2) operational definitions. Development of thinking in terms of processes and systems was just as vital. Two fundamental features of this thinking appear as early as pages 3 and 4 in *Out of the Crisis*. On page 3 is the "chain reaction" which, as we saw in Chapter 3, shows the simple logical progression by which the improvement of quality, in the broad sense of that word used in

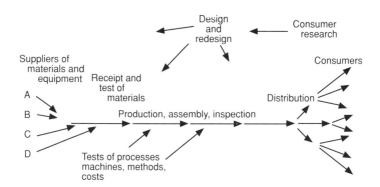

FIGURE 23
Production Viewed as a System
(Dr. Deming, *Out of the Crisis,* page 4)

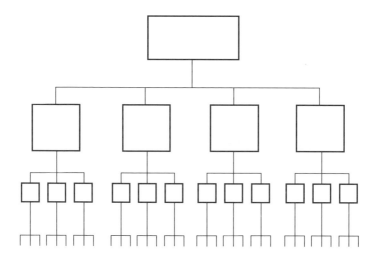

FIGURE 24
Hierarchical Structure of an Organisation

this book, leads also to improvement of productivity, and hence to survival, success, and expansion of the business. On page 4 of *Out of the Crisis* is the well-known diagram: "Production Viewed as a System," reproduced here as Figure 23. A version of this appears (also extremely early: on page 3 of the Introduction) in the published version of Deming's early lectures in Japan.

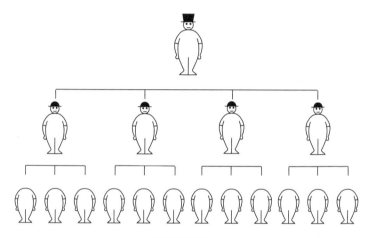

FIGURE 25
Cartoon Version of a Hierarchical Structure
(MANS Association)

One of the huge challenges of the Deming transformation is the organisation of management to reflect the unarguable logic of Figure 23, as opposed to simply using all that is implied in the traditional hierarchical structure of an organisation. The hierarchial structure is usually indicated by sketches such as Figure 24 or, amusingly but pointedly, in the wonderful cartoon[6] from the MANS Association shown in Figure 25.

The implication of the hierarchy diagram is that the most important customer of anything which goes on is the imme-

---

[6] Opinions differ as to whether those in the bottom row are headless or just bending over!

diate superior of the individual or group concerned. In fact, of course, the most important customer *should* be the one who receives the fruits of the labours of that individual or group, i.e. whose work is helped or hindered by them. Information, feedback, communication of any kind—exactly what is needed to aid improvement efforts—is indirect and inefficient unless some strong horizontal lines are drawn through the vertical structure. Those who should be the real internal suppliers and customers are otherwise effectively far apart from each other, whether or not they are physically so. And note that the *external* suppliers and customers don't even appear explicitly in the hierarchy diagram. Doesn't that leave the picture rather incomplete? As we have seen, the identification of suppliers and customers can hardly be taken for granted. For example, who is the customer of a hospital? The patient? The patient's family? The Doctor? The Head Nurse? The Ministry of Health? Society?

Use of Figure 23 enables Deming to immediately concentrate on the importance of the customer:

> "The consumer is the most important part of the production line. Quality should be aimed at the needs of the consumer, present and future."

(*Out of the Crisis*, page 5). Maybe the language in 1950 was not quite so direct, but it wasn't far off (this is again taken from the documented version of Deming's lectures in Japan):

> "The consumer is more important than raw material. It is usually easier to replace the supplier of raw material with another one than it is to find a new consumer. And a non-consumer, one who has not yet tried your product, is still more important to you, because he represents *a possible additional user for your product*."

Deming then spoke at length on *consumer research*, which he took care to illustrate in terms of *two-way communication* between manufacturer and both actual and potential customers. And this was still in the *Introduction* to his series of lectures!

Deming is unambiguous about the overriding importance of the customer:

"Who determines quality?
The customer does: he can decide what he buys."

Deming further talks in terms of quality being that which "entices" and "appetises" the customer. And the same sentiments hold in our extended notion of what we mean by "customer." The situation is, of course, somewhat different internally: our internal customer may not have the choice as to whether or not he deals with what we supply, but if the quality of what we supply is enticing and appetising to him, he will surely have greater pleasure and pride in what he can in turn supply to his internal customer, while the whole quality of our own work is similarly affected by what is supplied to us:

"People on a job are often handicapped
by inherited defects and mistakes."

Awareness of the substantial and involved interleaving customer-supplier structure makes us begin to realise how serious any mistake or any poorly-designed part of the system can be:

"A problem in any operation
may affect all that happens thereafter."

In other words, any supplier affects *all* subsequent customers. That concept is taken up impressively by Myron Tribus in his paper: "The Germ Theory of Management."

Concentration on, and sometimes the language of, damage

"downstream" caused by problems "upstream" occurs several times in *Out of the Crisis*. Problems due to multiple suppliers are a case in point, whether or not their separate inputs to the production system come in batches (*Out of the Crisis*, page 34) or are mixed (*Out of the Crisis*, page 355). Deming has remarked that there are beautiful mathematical methods for dividing up resultant variation into the components coming from the different suppliers. But the practical gains will not be made by hypothesising about what is happening upstream: they will be made by getting there and sorting out the problems in the upstream system, and thus preventing the occurrence of the downstream difficulties. Deming is fond of quoting Bill Scherkenbach:

> "Search upstream provides powerful leverage toward improvement of a mixture."

(*Out of the Crisis*, page 355).

As the reader will realise, this is one of the many lessons to learn from both the Red Beads and the Funnel Experiments. They show clearly the uselessness of, and harm caused by, trying to manage results rather than improving the upstream sources, i.e. the system which is producing the results. All the examples in Chapter 5 are relevant in this context. Some of the many other specific examples in *Out of the Crisis* emphasising the same message are the story from Reimer Express (pages 194–195), which includes the case of the driver who was found to be out of radio contact with the company's headquarters because of the mountains around Vancouver, the discussion of errors being caused in a sorting operation (pages 258–261) because pigeon-holes were beyond the reach of some of the ladies who were trying their best to do the job, and the several illustrations of dangerously-misleading roadsigns (pages 479–484), mostly from Deming's own observations in Washington. As he says:

"Costs are not causes—they come from causes."

The same is true of accidents and bad results of any kind. Why is it that governments, protesting mightily about crime, vandalism, football hooliganism, drugs, litter, and other manifestations of unhappy and disturbed people, don't start wondering a little more about the *causes* of those problems with the aim of prevention at source, rather than only attempting their suppression through fear of heavy penalties and punishment?

Chapter 9

THE DEMING CYCLE

The Deming Cycle guides us toward improvement. It is also known as the Shewhart Cycle, the PDCA Cycle, and the PDSA Cycle. Deming refers to it as the Shewhart Cycle since the concept seems to have its origins in Shewhart's 1939 book. The Japanese normally refer to it as the Deming Cycle, and that is what we shall call it here. Concerning the mnemonics, PDCA (Plan-Do-Check-Act) is the more common version, though Deming prefers PDSA (Plan-Do-Study-Act).

Shewhart's book began (yes, page 1) by separating out three stages of a quality control process:

1. The *Specification* of what is wanted;
2. The *Production* of things aimed at satisfying the Specification; and
3. The *Inspection* of the things produced to see if they satisfy the Specification.

On pages 44–45, Shewhart points out how differently this sequence needs approaching in this world where all processes are subject to variation, as opposed to another world which is comprised of exact science. In that other world, which unfortunately some people confuse with this one, the three steps would be independent of each other. As Shewhart says:

> "One could specify what he wanted, someone else could take this specification as a guide and make the thing, and an inspector or quality judge could measure the thing to see if it met specifications. A beautifully simple picture!"

OLD

FIGURE 26

The Quality Control Process Shown as a Line

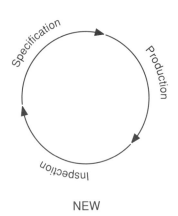

NEW

FIGURE 27

The Quality Control Process Shown As a Circle

(Figures 26 and 27 from Dr. Shewhart, *Statistical Method From the Viewpoint of Quality Control*, page 45)

140

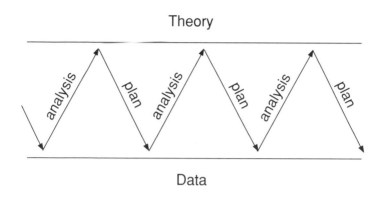

FIGURE 28
The Scientific Improvement Process
(Professor George E. P. Box)

Shewhart transformed the line in Figure 26 into a circle (see Figure 27), and referred to this circle as constituting a *"dynamic scientific process of acquiring knowledge."* After the first loop, much can be learned from the Inspection to guide improved Specification of what is needed. The Production process is amended accordingly, and the new output is inspected. This clarifies still further desirable improvements, and the circle continues.

The three stages in this circle, Specification-Production-Inspection, are not far removed from the first three stages in the PDCA or PDSA Cycle. Indeed, the version of the cycle shown on page 88 of *Out of the Crisis*, though expressed as four stages, has third and fourth stages which are, in effect, a division of "Inspection" (i.e. "Check" or "Study") into two steps which we may call Observation and Analysis, rather than incorporating the

"Act" stage. As with Shewhart's circle, Deming's Cycle is drawn in a way which makes clear that the sequence of steps may be repeated—in an improved manner, of course—using the knowledge gained in the previous loops. Professor George E. P. Box's[1] view of the scientific improvement process, as shown in Figure 28, is closely tied in with Shewhart's circle and this version of the Deming Cycle.

Deming now prefers a simplified picture of the Cycle which does have "Act" as its fourth step. Figure 29 is a copy of his current version, as he drew it during a seminar. As so often the case, he gives a very comprehensive description in very few words. Two particular points worth emphasising are that:

1. Deming recommends that Step 2 (usually referred to as "Do," though not on this copy) be carried out on a small scale—large enough to gain useful information, but no larger than necessary in case things go wrong; and

2. Step 4 ("Act") *may* be followed by another trip round the circle, incorporating the knowledge learned or with deliberately changed conditions in order to learn more, or alternatively it may be the final step, viz. the decision to adopt or abandon the Plan.

An aspect of Deming's work with which we are becoming familiar is his ability to focus on something which clearly makes a great deal of sense, yet which we mostly tend not to follow. Who could deny that the Deming Cycle summarises a wholly logical way forward for change of just about any kind? Yet do we use it when embarking on new activities? And have we used it in the past when trying out things for the first time which have

---

[1] George Box was for many years Chairman of the Department of Statistics at the University of Wisconsin in Madison, where I was privileged to spend a year immediately after obtaining my doctorate in 1967. George is now Research Director at the Center for Quality and Productivity Improvement, also in Madison.

now become part of the way we live, work, or play? I think not. How much of what you have done today has been carried out in the way you learned it, maybe years ago, without a thought of possible improvement?

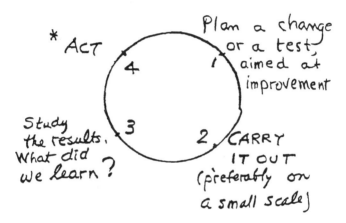

THE SHEWHART CYCLE

Plan a change or a test, aimed at improvement

*ACT

4

1

Study the results. What did we learn?

3

2 CARRY IT OUT (preferably on a small scale)

* ACT. Adopt the change.
or   Abandon it.
or   Run through the cycle again, possibly under different environmental conditions.

FIGURE 29
The Shewhart Cycle
(Drawn by Dr. Deming)

143

My BDA colleague Ian Graham,[2] with whom I am proud to have shared many presentations on Dr. Deming's work, compares what *should* happen with what most usually *does* happen in Figures 30 and 31.

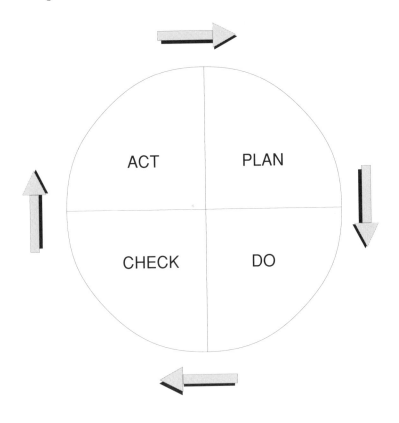

FIGURE 30

The Deming Cycle

(Designed by Ian Graham for comparison with Figure 31)

---

[2] Ian Graham made a substantial contribution to the *Doctor's Orders* video (first screened by Central ITV on 18 October 1988); at the time he was Director of Quality at Hewlett-Packard, and he is now a member of PRISM Consultancy.

Why does Figure 31 happen rather than Figure 30?   I guess it's because we have all been brought up to "Do." Doing is "productive," whereas Planning, Checking, and Studying are "nonproductive." By Doing we feel we are getting somewhere, whereas while Planning—thinking, talking, studying—we feel we haven't yet made a start. There is strong influence here from our results-oriented society—one can easily produce some measures of what has been Done, but not so easily of what has been Planned.

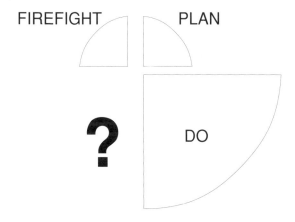

FIGURE 31
The Usual Approach to Planning
(Ian Graham)

The results-oriented society also sets great pressure on our being right all the time—or, seemingly more important, *appearing* to be right. How often do we hear a politician admit or confess (you see—even those words are pejorative) that he has been wrong? Yet it seems remarkable that two or more parties, generally contradicting each other, can all be right all the time. And, if a politician does say he's been wrong about something, the media come down on him like the metaphorical ton of bricks! Politicians are apparently respected, if grudgingly, for sticking to their guns even by people who think they are wrong.

Also, because we are so used to living life according to Ian Graham's second picture rather than his first, fire-fighting becomes a highly-respected activity—indeed, it's the way that a lot of people win promotions. We forget that it would be better if the fires didn't get set in the first place.

The notion of circles of improvement is seen in various guises in Deming's work. The most obvious illustration is the "Production Viewed as a System" flow-diagram (Figure 23), which we discussed in the previous chapter. Every piece of feedback is for Checking and Studying and Acting appropriately, and Planning for yet more improvement. A further manifestation is seen on page 180 of *Out of the Crisis* in diagrams more than reminiscent of the ones shown earlier from Shewhart's book. There, the old way of manufacturing is pictured as in Figure 32, while the new way is shown as a circle as in Figure 33. The four steps in the circle are:

1. Design the product;

2. Make it; test it in the production line and in the laboratory;

3. Put it on the market; and

4. Test it in service; find out what the user thinks of it, and why the non-user has not bought it.

And, of course, Step 4 leads into a new Step 1: redesigning the product, and the circle begins again.

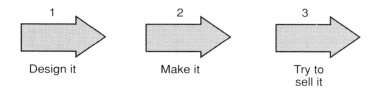

FIGURE 32
The Old Way of Manufacturing

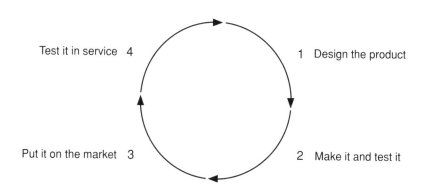

FIGURE 33
The New Way of Manufacturing

(Figures 32 and 33 adapted from Dr. Deming,
Out of the Crisis, page 180)

A further example of this circle-thinking comes when Deming discusses his statement: "Experience teaches nothing unless studied with the aid of theory" (see Chapter 16):

> *"Experience teaches us*
> *(enables us to plan, to predict)*
> *only when we use it*
> *to modify and understand theory."*

Concentration on Planning is, without doubt, a feature which distinguishes Japanese practice from our own. On page 85 of *The Deming Route to Quality and Productivity*, Bill Scherkenbach first quotes from the January 1985 issue of *Automotive Industries*:

> "In typically Japanese fashion, NUMMI's[3] Fremont plant will have an extremely slow startup. In the first few weeks, the sole line ran only an hour or two a day. It will take the better part of six to nine months before full production is reached; extensive training will take place during this time."

and then comments:

> "Many people in management would say that the above example is bad business, that with all the machinery installed, they have yet to begin to generate a return on their assets."

Five years later, the success of the Fremont plant is well-established, to say the least.

Sometime ago, I was talking with a Nissan salesman about the Japanese way of doing business and, in effect, the Deming Cycle entered into our conversation. Eventually, he said that our discussion had clarified in his mind something which had been puzzling him for years. He had previously been a salesman

---

[3] NUMMI: New United Motor Manufacturing Inc.: a joint venture between General Motors and Toyota to find out if Toyota's manufacturing techniques could be successfully implemented with a unionised work-force and American suppliers. They could. They can.

for British Leyland. Suddenly, during that time, Nissan started buying up all the used mini-cars that they could lay their hands on, including wrecks, and shipped them back to Japan. According to this salesman, the numbers involved were not measurable in, say, dozens but in *thousands*. He had never understood why. I asked him what he thought of the Micra—Nissan's mini-car, which came onto the market about four years after all those old cars had been sent to Japan. He knew that the Micra was one of the leading cars available in its class. Now he knew why. Detailed analysis of those shiploads had been part of the Planning process.

Chapter 10

FAILURES WITH FIGURES

An apparent paradox in Deming's approach to management, originating as it does with *statistical* thinking, is that he lays great stress on the *limitations* of figures. Whereas to most people the word "statistics" means nothing more nor less than figures, we are already familiar with Lloyd Nelson's highly perceptive statement of which Deming is so fond:

> "The most important figures needed for management of any organisation are unknown and unknowable."

And, even if our figures are not unknown, Deming reminds us of Professor John Tukey's warning that:

> "The more you know what is wrong with a figure, the more useful it becomes."

(As a help toward this, the evaluations of the speed of light mentioned in Chapter 7 did indicate *ranges* of values as opposed to explicit values.) Deming himself, referring to the highly misleading figures on biggest export-dollar earners (Chapter 1), says:

> *"Figures are only useful*
> *if we know how to use them;*
> *they are no use on their own."*

And M. J. Moroney, author of *Facts from Figures* (the only statistics book that many people have ever read), warns us:

> "The words 'figure' and 'fictitious' both derive
> from the same Latin root: 'fingere.' Beware!"

I could go on. Instead, if you would like to continue with this apparent diversion, I refer you to John Bibby's amusing little booklet: *Quotes, Damned Quotes, and ...* .

The originators of all of the above quotes, including Deming himself of course, are *statisticians*. What are they trying to do? Put themselves out of business? Some might well interpret Deming's statement from Chapter 7 that "There is no true value of anything" in that way; and likewise, from Chapter 18: "There is no such thing as a fact concerning an empirical observation."

Some management people seem to have become obsessed with numbers: "Don't bother me unless you can show me the figures!" In these days of supposed quality revolution, many industrial statisticians are told that their job is to "show me how to measure quality" or " ... productivity." Unfortunately:

> "Measures of productivity are like statistics on
> accidents: they tell you all about the number of
> accidents in the home, on the road, and at the
> work-place, but they do not tell you how to reduce
> the frequency of accidents."

(*Out of the Crisis*, page 15). Another quotation often bandied about is: "If you cannot measure it, you cannot manage it." Deming is pretty summary about his dismissal of that one: "Totally wrong—nonsense!"

The origin of the eighth of the 14 Points (Drive Out Fear) was Deming's realisation that fear guarantees the presentation of wrong figures (see Chapter 26). Figures, like fear, have in many cases become a *weapon* of conventional management. Indeed, figures are often used to *generate* fear, particularly through some of Deming's strongest abominations of bad management: MBO, arbitrary goals and targets, merit rating, and grading in schools.

Whereas some management seem to pin their faith on figures, the layman has some considerable scepticism about statistics, and thus (I suppose) by implication, statisticians. Hence the old saying about proving anything with statistics, and the famous quote about the three kinds of lies.[1] That scepticism, which is often not at all unjustified, immediately lends weight to the need for wide understanding and use of operational definitions which, as *le mot juste*, must rid us of the ambiguities (whether intentional or not) leading to such mistrust. Remember that Deming believes Shewhart's emphasis on operational definitions and his own concentration on the same topic in his early lectures to the Japanese were at least as important as their work on variation and control charts.

The purpose of this chapter is to warn the reader against falling into common traps. These include paying too little attention to matters for which figures are unavailable—or alternatively forcing out some inappropriate figures if "real" ones cannot be obtained, the misuse of figures for purposes for which they are unsuited, faulty analysis of figures, and misunderstanding and false use of averages.

Why *did* Lloyd Nelson make that statement about unknown and unknowable figures, which seems so strange to people

---

[1] "There are three kinds of lies: lies, damned lies, and statistics"—usually attributed to Benjamin Disraeli but, according to John Bibby, more likely to have originated from Mark Twain.

hearing it for the first time? What can he mean? Deming answers on page 122 of *Out of the Crisis* with a long list of crucial matters which can never be quantified yet which are more important to success or failure of the business than virtually anything which can. They include the value of having happy customers, of improvement upstream, of faith in management's principles and aims, of teamwork for improvement and innovation inside and outside the company; and the cost of having unhappy customers, and of merit rating, and other inhibitors to pride of workmanship.

Nowadays he would add to the list the huge costs of suboptimisation rather than optimisation, failure to understand leadership, and use of the Rules of the Funnel: tampering (Rule 2), and worker training worker, and executives meeting together, trying to improve quality, market, and profit, without Profound Knowledge (Rule 4). Such costs are not only unknown and unknowable: "they are unrecognised and not even under suspicion." Management who play with ideas for quality and productivity improvement without knowledge and understanding expect to be given figures on what the gain will be. But:

> "He that expects to quantify in dollars the gains that will accrue to a company year by year for a program of improvement of quality by principles expounded in this book will suffer delusion. He should know before he starts that he will be able to quantify only a trivial part of the gain."

(*Out of the Crisis*, page 123).

Even when we have figures, misuse of them is rife. How many are the decisions made on the basis of comparing just two figures—when, if the underlying process is in statistical control, there is a 50–50 chance as to which one is the larger? One way of explaining Shewhart's fundamental work of over 60 years ago is that it provides help in deciding when action of a particular type is *not* justified (in particular, treating common cause variation as though it were due to special causes). Even if some people in management progress to looking at a record over time, rather than just comparing a figure with its forecast, or with budget, or with last month's figure, or with last year's figure, their eyes still almost automatically look for the high and low points, without seeming to understand that *any* sequence of values must have a highest and must have a lowest—including sequences produced by random variation. Yet is that difficult?

The concentration on understanding *variation*, necessary for any justifiable analysis of figures, has shifted us away from the traditional emphasis on *averages*. Most people will agree that they have never studied anything about the theory of variation, although they are quite at home with averages. Are they? There are plenty of examples on pages 56–59 of *Out of the Crisis* which will dispute that.

Heero Hacquebord elaborated to me on the events reported on page 59 of *Out of the Crisis*, where a teacher warned him that his young daughter was "below average" in both of two tests. He told me of the great distress caused to his daughter who, I gather, was in fact neither then nor now noticeably dim! Surely, under the simplest-possible probability model, one in four children will have been below average on both tests purely through chance and through no fault of their own. Hardly good grounds, surely, for either scaring parents or demoralising children. In fact, Heero's daughter did feel humiliated and depressed by the experience. She had to be given help to get over it. She recovered. Suppose she hadn't? "A life lost or, at least, seriously harmed," concludes Deming. As he agrees, the teacher had the best of intentions: she was only doing her best. And we immediately recall the warning that good intentions and best efforts are not enough—we need to know what to do, and why.

The New Zealand Government used to have a national school-leaving examination, so regulated that half of the entrants passed and half failed (again see page 59 in *Out of the Crisis*). The consequences to the future careers and lives of those students sent out of school are easy to imagine. At a seminar there, Deming talked about the matter. The Government abolished the scheme.

We know Deming's opinion of performance appraisals and merit ratings, of which a particular case is gradings in school; we shall read more in Chapters 30 and 31. But such schemes become even more of a cruel mockery if they are subject to a forced distribution. The New Zealand examination was forced into a very simple distribution: half had to pass, the other half had to fail. Even if New Zealanders were a race apart, with intelligence and ability unmatched anywhere on the rest of the globe, still half the students had to fail. And the illogicality and unfairness is not restricted to 50–50 averages:

"Some companies take the top 10% of the class— serve them right!"

On a lighter note, Myron Tribus tells a great story about a vote in the Trades Union Congress in Britain on the motion that no wage paid in the country should be below average. It apparently failed by just three votes. As Deming wryly comments: "How I wish it had succeeded!" Perhaps the TUC had been listening to the Australian Minister of Labour who, according to John Bibby, said even more ambitiously in 1973:

> "We look forward to the day when everybody will receive more than the average wage."

And, as a final thought about averages, I cannot resist the ancient quote from *Punch:*

"The figure of 2.2 children per adult female was
felt to be in some respects absurd, and a Royal
Commission suggested that the middle classes be
paid money to increase the average to a rounder
and more convenient number."

Chapter 11

IS CONFORMANCE TO SPECIFICATIONS
GOOD ENOUGH ?
THE TAGUCHI LOSS FUNCTION

We commence this chapter with another highly illuminating story from the Ford Motor Company.[1] The story is important because it caused a breakthrough in thinking in the minds of
many Ford personnel—one of the many breakthroughs needed *en
route* from traditional thinking to the new philosophy. That
breakthrough is the understanding that it is no longer good
enough to consider quality (even in the limited sense of measuring characteristics of a product or service) merely in terms of conformance to specifications.

Ford's 1983 ATX transaxles were obtained from two
sources. The Ford Transmission Plant in Batavia, Ohio produced
most of them, and the others were manufactured by a Mazda
plant in Japan. Even though the Batavia and Mazda transaxles
were produced with reference to the same blueprints, customer
feedback made it clear that there were definite differences
between the two products. Customers with the Mazda-made
transaxles expressed greater satisfaction with their vehicles,

---

[1] I am most grateful to Don Wheeler for filling me in with many details
concerning this story of which I was previously unaware.

and the proportion of warranty claims for the Mazda-made transaxles was much less than that of the Batavia transaxles.

Because of these differences, Ford conducted a detailed study of ten Batavia transaxles and ten from Mazda. Each transaxle was evaluated on the test stands, prior to being disassembled. Literally every performance characteristic and every physical characteristic which had a specification was measured. The results looked good: all twenty transaxles were found to conform to all the specifications.

However, histograms which were constructed of the measured characteristics from the two samples were not at all similar. The Ford measurements were often spread out over almost the whole of their specification ranges. Even some of the more critical dimensions had histograms which covered over 70% of their specification ranges. In contrast to this, the histograms of the various characteristics from the Mazda-made transaxles were generally concentrated within the middle 25% of the specification ranges, while some of the more critical dimensions showed no discernable variation at all.

The video[2] which Ford made of this story tells about one set of these critical dimensions: the bore diameters in the valve body.

These bore diameters were checked by the regular floor inspector using a sophisticated air-electronic gauge which measured to the nearest ten-thousandth of an inch. A valve body would be checked by placing each bore, in turn, on a probe and then rotating the part. The rotation allowed the gauge to measure the bore diameter in all directions, and this was necessary since, of course, nobody can manufacture perfectly round holes. In order for a bore to be classified as satisfactory, not only did it have to meet specifications, but also the difference between the maximum and minimum diameters had to be less than a specific

---

[2] *Continuous Improvement in Quality and Productivity*

value. In spite of the variation observed, all of the bores on all of the Batavia valve bodies were classified as satisfactory.

The inspector became puzzled, however, when he began to examine the Mazda-made valve bodies. As he rotated the first part in order to measure the diameter of a bore, the reading on the gauge did not change. Mystified, he tried the second Mazda part. The same thing happened. And then the third. And then the fourth. Naturally, the inspector was by now convinced that the gauge was no longer working correctly, and so he called in the gauge manufacturer to remedy the fault. As the reader may have guessed, the repairman found nothing wrong—which, of course, was confirmed when they measured the Batavia parts again. Quite simply, to the nearest ten-thousandth of an inch, the bores in the Mazda-made parts were (a) round and (b) absolutely identical from part to part. All of this led John Betti, who was then the Vice-President of Powertrain and Chassis Operations, to say:

> "While we've been making great progress in meeting my original objective of building to print, our not-so-friendly competition was making great strides in building uniform parts—every part just like the part ahead of it, and just like the part following, with very little variation. While we were arguing about how good the parts had to be, they were working hard on making them all the same. *We* worried about specifications; *they* worried about uniformity. While *we* were satisfied and proud if we were to print, and then worried about keeping it to print, *they* started with the part to print, and worked on continuous improvement in the uniformity of the parts. Control, uniformity, continuous improvement."

Deming also briefly quotes John Betti's remarks on page 49 of *Out of the Crisis.*

It surely follows that conformance to specifications is not a sufficient criterion to judge quality. Indeed, to attempt to do so is an immediate contradiction to the insistence on continual improvement which is one of the fundamentals of the Deming philosophy. For the latter is always looking for better quality, whereas the former provides no encouragement to do better once specifications are being consistently met. On the contrary, the basic ethic of some other attitudes to quality, including considerations of "Cost of Quality," tend to mitigate against further improvement, arguing that customer requirements have been met and so it is not worth expending any more time, effort, or money on that particular process.

Such attitudes are in contradiction to the Deming chain reaction (see Chapter 3). Clearly, they are also not shared by the Japanese, whose processes have often been developed to the stage where measured quality-characteristics only cover the middle third or even fifth of the specification range (see *A Japanese Control Chart* for an example of this). But why? It *must* cost more time and money to reach such excellence, mustn't it? So presumably there are spin-offs. What are they?

First, as we have seen in the Batavia story, there is enhanced reputation in the eyes of the customer, which naturally tends to attract extra business—this is the Deming chain reaction in operation. But there are many others. The work carried out to improve processes to such a degree brings with it knowledge which can often be used in other processes and operations. And that knowledge gets the processes into such good shape, and makes them so well-understood, that the chance of anything ever going seriously wrong with them is negligible—which in itself constitutes huge savings.

It also contributes to innovation (see Chapter 14)—not only is more time freed up for research and development, but the time to put such research and development into operation is reduced because the technical help is so much more advanced.

Operations become smooth-running, "hassle-free." Even when a process goes out of statistical control, and the problem cannot be sorted out quickly and easily, production can often carry on as normal—for if control limits are very close together, so that the process is performing well within specification limits, it is quite possible that a move out of control will still not send anything even near those specification limits. That is not to say the problem shouldn't be tackled: it should, and as soon as possible. The point is that, if the problem is a difficult one to solve, there is no need to shut down the production line until it is sorted out—as *is* the case if the control limits are further apart.

Many observers at Japanese factories have remarked not on massive automation and other wizardry—in fact, they often see no more of that than they are used to at home—but on the continuous, smooth, uninterrupted flow of production, which *is* very different from their normal experience.

Further spin-offs (non-quantifiable, of course) lie in the enhanced morale and pride of workmanship in developing a superb product or service and, having developed it, working with it: truly, joy in work (see Chapter 13).

Finally, there is minimal further cost related to product or service once it has been received by the customer—i.e. minimal rework and warranty costs. The point that improved quality results in less rework is emphasised on the very first page of *Out of the Crisis*, though there it is in the context of showing why improved quality results in increased productivity.

Our discussion so far in this chapter has really been more in support of the general philosophy of continual improvement rather than with the particular problems of specifications. But managing by conformance to specifications brings its own problems. This is not to say that specifications have not served a useful purpose over the years. Deming himself pointed out at one seminar that specifications had been very good for North American industry at times when both their quality and that elsewhere in the world had been pretty wretched: specifications helped make things which were good enough to sell the world over. But:

"Not now.
Now, others have got ahead.
On everything?
No, not everything.
But on most things which matter."

And:

"Specifications are not wrong.
They are just not sufficient."

(This is also his attitude to many of the common ideas about how quality can be achieved—see Chapter 17.)

If we go a long way back, specifications were not necessary. That was in the days before mass production, when components could be individually crafted to fit each other. But the advent of mass production did away with that possibility. What was the alternative? It would be nice if one could state the ideal nominal value and have everything made to that. But this is a world full of variation, and life is just not that easy.

The almost automatic solution is to define an interval around the nominal, the ends of that interval being the *specification limits*. Items whose measurements lie within the interval, i.e. between the specification limits, are taken as acceptable, but those failing to do so are rejected. Of course, this serves a useful purpose. It ensures that measurements close to nominal are accepted, while those far away from nominal are rejected. And, naturally, that's fine—as far as it goes.

But now let's study some of the problems which specification limits can cause. To take a common and easily-understood example, consider the manufacture of shafts and cylindrical sockets into which the shafts are supposed to fit comfortably—not too tight and not too loose.[3] There is a difficulty here which we shall not discuss: that concerns how constant the diameters of the shafts and sockets are along their lengths, and indeed (remembering the Batavia example again) how round they are. So, deciding how to measure the diameters is no trivial task and is in clear need of an operational definition (see Chapter 7). For our purposes here, we will suppose that variation along the lengths of both sockets and shafts is markedly less than the variation between the pieces.

Let us consider some of the problems which may occur if the fit is not ideal. If the fit is on the tight side, extra friction will be experienced when the machine is running. To overcome this friction, extra power or fuel has to be used, incurring higher running costs. There could also be some localised overheating, leading eventually to some deformation and inefficient running. If the fit is too loose, lubricant may be lost, perhaps leaking out to cause damage elsewhere. At the very least, replacing the lubricant may prove expensive, both because of the cost of the lubricant itself and through having to stop the machine more frequently to carry out maintenance. A loose fit will result in vibration, causing noise and oscillating-stress which is likely to lead to a shorter working life through fatigue-failure. Generally, such losses will increase progressively according to the badness of the fit. Some such losses will therefore be incurred even when both parts are within whatever specifications are laid down.

Suppose that the socket diameter is nominally 13.25 mm and, for the moment, let's presume that "perfect" sockets are being made, i.e. with diameters exactly equal to 13.25 mm, at least as far as we are able to measure. (This is unrealistic and we shall relax this assumption before long.) Consider therefore the diameter of the shaft. What should it be? Not 13.25 mm exactly,

---

[3] I am grateful to Malcolm Gall for his help with this illustration.

for then it wouldn't slide into the sockets—the fit would be too tight. Maybe 13.15 mm would be good, with the 0.10 mm gap being filled with lubricant. Well, that may be ideal—but what is practical? Let's go for some specification limits; for the sake of argument, let's take them in the form of 13.15 mm ± so much. What should "so much" be? It couldn't be as high as 0.10 mm, giving 13.05 mm–13.25 mm, because we have already pointed out that a shaft of diameter 13.25 mm won't fit. Let's try 13.15 ± 0.08 mm, i.e. 13.07 mm–13.23 mm. That may still give too much leeway: 13.23 mm would produce a very tight fit, and 13.07 mm might be a rather loose fit.

So now the argument starts. These specifications are usable but they cause difficulties. We'd be much happier with 13.15 ± 0.05 mm, i.e. 13.10 mm–13.20 mm. The trouble is that the Production Department tells us it can't work to such tight tolerances. However, everybody knows that Production is always covering its you-know-what (CYA) in this way! On the other hand, Production knows that the design engineers always ask for more precision than is needed. To quote John Betti from the Ford video again:

> "Everybody knows that the engineer asks for tolerances twice as tight as he needs because chances are the manufacturing guy will build it to print only part of the time at best."

Arguments continue, perhaps finishing with a compromise of 13.15 ± 0.07 mm. Incidentally, this very common type of "negotiation" is an indication of the traditional inner-competitive environment, rather than the Cooperation: Win-Win culture which is of top priority in Deming's current teachings (see Chapter 15).

Whatever we finish up with in terms of specifications, we begin to see some illogicalities in the whole concept. In no real sense is everything within the specifications "good," and everything outside the specifications "bad." The shafts having diameters near the nominal 13.15 mm are fine, but those near the specification limits of 13.08 mm and 13.22 mm are less good for obviously different reasons: those near 13.08 mm fit rather loosely, and those near 13.22 mm rather tightly. The same would have been true, of course, but to a lesser extent, had we been able to persuade Production to agree to 13.15 ± 0.05 mm. Furthermore, whichever specification limits we use, it is surely quite illogical to say that a shaft with diameter *just* inside specifications is *good* and should be accepted, while a shaft with diameter *just* outside specifications is *bad* and should be rejected. For example, in practice are we even going to be able to tell any difference between shafts with diameters 13.075 mm and 13.085 mm, or 13.079 mm and 13.081 mm?

Now, all this discussion has assumed we are producing perfect sockets of diameter 13.25 mm. But we can't do that. The process for production of sockets is also subject to variation—perhaps even more so than for the manufacture of shafts, for it may be a more complicated process. So, off we go again.

Production is adamant that it can't work to better than 0.20 mm in the manufacture of sockets. Is that really true? Who knows? Probably, as before (in the *competitive* rather than cooperative environment), they will be covering themselves with a safety-margin. As implied earlier, it is quite common for Production to quote as their limit of precision at least double what they can actually manage. However, after tough bargaining, they won't come down any further than saying they will try 13.25 ± 0.15 mm, i.e. 13.10 mm–13.40 mm. Of course, this sends all of our earlier deliberations out of the window—clearly, many shafts within the specification range of 13.08 mm–13.22 mm will now stand no chance of fitting the sockets with which they get paired!

And so, on goes the bargaining again. Maybe Production is eventually forced down to 13.25 ± 0.10 mm, i.e. 13.15 mm–13.35 mm (which is what it thought it could do anyway—the double-up strategy has worked again!). So, how about specifications for the shafts? It's now pretty hard to even decide on the nominal value, never mind the specification limits. The 13.15 mm won't do— even that will not fit into sockets near their lower specification, let alone anything a little larger. Maybe, after more fighting, we get Production to agree to 13.05 ± 0.05 mm: 13.00 mm–13.10 mm. This certainly avoids non-fits (as long as both shafts and sockets are within their respective specifications). But, with this arrangement, there are going to be a lot of undesirably *loose* fittings. Maybe we should slightly raise the specification interval for the shafts, risking a minority of non-fits for the benefit of many more good fits. And so the frustration continues.

What surely becomes clear is that specification limits are a poor way indeed to define and control a process. And, of course, most manufacturing processes have a considerable number of similar interlinks, at *all* of which such debate can arise. Almost always, there is no interval which realistically divides product or service into good and bad. Use of specifications gives the impression that there is. But, in reality, what happens is that specification limits are laid down somehow or other, and "good" and "bad" become defined in terms of them—we have then entered the realm of the self-fulfilling prophecy.

Of course, several products and services effectively use specifications in their description, e.g. 18 carat gold or two-hour film processing. Apart from such examples where the importance of specifications is self-made, Deming suggests just one case where a specification limit represents a real boundary between good and bad. His example is the proportion of the rare earth element columbium in sheet steel, which is apparently quite critical as regards the steel's welding properties. One might be able to think of a few other similar examples, but surely they form a very small minority.

It will be wise here to clear up a point of confusion which

sometimes arises. The use of specification limits, about which we are now less than enthusiastic, may seem very similar to the use of operational definitions (Chapter 7), whose necessity we have stressed. Indeed, if specification limits are needed, *they* should of course be operationally defined—including an unambiguous description of the method of measurement. There has to be a rule defining when something will be held back, reworked, retyped, scrapped, apologised for, etc., else confusion will reign. But that is very different from regarding this rule as specifying what is *bad* and proceeding as if everything else not covered by the rule is *totally satisfactory.* Of course we don't want to produce, let alone allow our customer to receive, anything which is *bad*—that should go without saying. The difference is the thinking about what is *not* bad:

"Of course, we don't wish to violate specifications, but we must do better."

Certainly, the Japanese quote specifications. But in Japan, specifications play more the role of an initial benchmark, whereas traditionally our aim, our end-point, has been to eventually reach them. The aim should be to design systems which will meet specifications from the start, and then to carry on improving from there. That is a very different philosophy, and it has very different effects.

We saw at the end of Chapter 7 that Deming sometimes says Shewhart's 3σ limits provide "an operational definition of a special cause" or rather "... of when to look for a special cause." We know well from Chapter 4 what he thinks of the practice of putting any other "action" lines on control charts—including specification limits. So that is a pertinent case indeed of where the use of an operational definition certainly has nothing at all to do with specification limits, and also where its use is central to the improvement effort itself.

So, if goodness and badness cannot be represented in terms of specification limits, what can we do instead? Let's return to the manufacture of those sockets of nominal diameter 13.25 mm. The truth of the matter is that, the closer the diameter is to the nominal, the better the socket is. And the further away it is, the worse it is. There are no sudden *steps* from good to bad, or from bad to worse. The seriousness of errors increases continuously as we move further and further away from the nominal. 13.25 mm is best—better than anything else. 13.26 mm is not quite so good; it's probably not going to cause much harm, but it is not precisely as good as 13.25 mm. Similarly, 13.27 mm is worse than 13.26 mm; and 13.28 mm is worse than 13.27 mm. And so on. We could work in terms of other numbers of decimal places also: so 13.274 mm is worse (just!) than 13.273 mm. There will be a similar pattern on the lower side of the nominal 13.25 mm—things may get worse at the same kind of rate as when we move upwards from 13.25 mm, or the rate may be faster or slower.

There is one further development to the argument which we need to make. We have said that it will not cause much harm for the diameter to be 13.26 mm instead of 13.25 mm—in fact, the harm will probably be pretty negligible. What about 13.27 mm instead of 13.26 mm? Is that also "pretty negligible"? And then 13.28 mm instead of 13.27 mm? The question is whether or not all 0.01 mm changes in the diameter have the same worsening-effect as each other. The answer is that, virtually always, they don't. The truth is that, the further we get away from nominal, the more serious each 0.01 mm move generally turns out to be. The next example will show this effect more clearly.

To some readers, especially those not engaged in manufacturing activities, this discussion may not seem immediately relevant to their own environment. But it is. Pete Jessup of Ford Motor Company has suggested an example which relates to every one of us. Consider the temperature of the room you are in. We're all different—some like it hot, some like it cool, so the "ideal" or "nominal" temperature will depend on the people involved. Maybe you are the sole occupant of an office. Or maybe there are several people in the same work-area. No problem. If we consider everybody's preferences, there'll be some figure which is best on average. Or suppose you are on your own. What temperature do you prefer? That is, where do you set the thermostat (if you are lucky enough to have one)?

Let's say the ideal temperature is 70°F. What if the room temperature is not quite equal to that? Suppose it's 71°F. That won't make a lot of difference. It will presumably make *some* difference, else we'd be contradicting our assumption that 70° is *best*. But not much. If it's 72°, then that could perhaps just be taking the edge off your comfort. What about 73°, or 74°, or 75°? By 75° you may be beginning to get noticeably uncomfortable: that 5° excess *is* affecting your work. But it is still describable in terms of inconvenience, causing only a minor loss of efficiency. A further 5° would be a rather different matter. At 80°, it's getting pretty difficult to concentrate. The sweat may just be beginning to roll. You're checking your watch far more often, looking forward to the time when you can get out of there. At 85° or, for the most resilient, 90°, it's pretty much impossible to do anything useful. Notice that each 5° increase has a bigger impact than the previous 5° increase.

A similar picture is easily seen on the down-side. A drop to 69° is presumably going to be hardly noticeable. 68° might just be. By 65° there is some discomfort. 60° is really causing trouble. And at 55° it is well-nigh impossible for anybody to sit in an office and do anything of much value. Again, each 5° drop is more serious than the one before. (The same would be true if we considered drops of, say, 6° or 3° or 2.5°.) Notice the parallelism with our final remarks about the increasing seriousness of errors in the socket and shaft diameters.

Now, recall where we started, i.e. attempting to define satisfactory quality by using specification limits. Would it not be absurd to try to describe this situation in terms of an interval (range) of "satisfactory temperatures," with everything outside that range being "unsatisfactory"? Assume we are restricting attention to intervals of the form $70° \pm$ so much. Do we say $68°$–$72°$, or $65°$–$75°$, or $66.5°$–$73.5°$? There is, of course, no "right" answer. I would suggest there isn't even a sensible answer, let alone a right one. Whichever interval you choose, say $65°$–$75°$, the facts remain that:

(a) temperatures near $65°$ or $75°$ are *not* as suitable as temperatures near $70°$; and

(b) you're not going to be able to tell the difference between $74.9°$ and $75.1°$, so it just doesn't make any sense to class $74.9°$ as satisfactory and $75.1°$ as unsatisfactory.

The type of argument in (b) will be true whatever interval we choose. The argument in (a) could be weakened by choosing a very narrow interval, say $69°$–$71°$, or $69.5°$–$70.5°$. But that raises more problems. First, if the interval represents specifications on your office heating or air-conditioning equipment, these specifications may be so tight that it is either impossible, or would be ridiculously expensive, to meet them. Secondly, temperatures outside such a narrow range, though not achieving optimum performance, are nevertheless nowhere near so bad as to be realistically described as "unsatisfactory."

Our arguments seem to be getting nowhere fast! However hard we try, we find that attempting to define acceptable standards of quality in terms of specifications—intervals—leads us towards illogicalities and impracticalities. Yet this *is* the way many industrialists *have* judged quality for decades.

For realism, a totally different approach is needed—one which does not require artificial definitions of right and wrong, defective and nondefective, conforming and nonconforming;

one which instead accepts that there is a "best" or "nominal" value, and that any divergence from this nominal causes some harm or difficulty, following the kind of pattern which we have described with respect to both the shaft and socket diameters and the temperatures. *Taguchi's loss function* does just this. It was described and discussed[4] in a paper which Genichi Taguchi read in Tokyo in September 1960; Deming was present. Shown graphically, Taguchi's loss function is usually depicted in a form such as that shown in Figure 34.

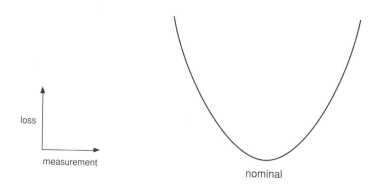

FIGURE 34
The Taguchi Loss Function

The value of the quality-characteristic is registered on the horizontal axis, and the vertical axis shows the "loss" or "harm" or "seriousness" attributable to the values of the quality

---

[4] It is not claimed that Taguchi *invented* the loss function, in particular the commonly-used quadratic loss function (see Chapter 12). Indeed, it has been used by mathematicians for at least the last two centuries; names such as Gauss, Laplace, and de Moivre come to mind. But Taguchi *has* done a great job of raising awareness of the need to apply this highly-important concept in the kinds of context being studied here.

characteristic. This loss is taken as zero when the quality-characteristic achieves its nominal value,[5] but is positive otherwise. However, reflecting the arguments which have been developed, very little loss is incurred while the quality-characteristic is fairly close to the nominal value. But, as the value moves away from the nominal, the loss increases at an ever-faster rate.

Now we should not claim that there is necessarily any "correct" choice of Taguchi loss function—especially following many comments in the previous chapter. How does one measure the loss due to an inaccuracy? Is it in terms of reduced quality of the end-product, possible rework, customer dissatisfaction, loss of morale in the work-force for handling inferior quality, etc.? Taguchi's own position on this question seems to have varied; on the one hand, he has referred to the loss as being "loss to society," while his experimental design methodology[6] requires a specific evaluation of loss. In either case, statisticians should note that this evaluation is *not* wholly their job; invariably, it will have to come substantially from real knowledge of the subject-matter. Teamwork between statisticians and subject-matter experts is another facet of the Cooperation: Win-Win environment.

Given that difficulty, what are the advantages of Taguchi's loss function over the use of specifications? Plenty.

First, although there may be debate on the exact nature of the loss function in any particular situation, that debate is at

---

[5] In absolute terms, there is no need for the loss at the nominal value to be zero: it could be positive, or it could be negative—hence representing a level of profit rather than loss. (Indeed, the development of the theory in terms of loss rather than profit might be regarded as unnecessarily pessimistic!) Some statisticians prefer the term "risk function" to "loss function," where risk is defined as the loss minus the minimum possible loss. For most purposes, the gain in simplicity provided by the risk function outweighs any possible advantage of retaining a more general loss function.

[6] The reader may know that Taguchi's approach to experimental design is being used quite widely in industry, though amidst controversy in the minds of some statisticians.

least based on a logical foundation as opposed to one which could never be a valid representation of reality.

Second, the Taguchi loss function always keeps in our minds the necessity for continual improvement—if there are discrepancies from nominal (and there *will* always be), then loss is being incurred, so the need for improvement (reduced variability) is ever-present. That is very different from the thought that 100% conformance to specifications is somehow the ultimate quality aim, which John Betti admits *was* his thinking prior to the events reported at the start of this chapter.

Third, even very rough assessments of loss functions yield highly useful information for prioritisation of improvement efforts. Even accepting the need, expressed in the fifth of the 14 Points, for continual improvement in all processes and systems, clearly we cannot do everything at once: we have to be practical. An order of priority has to be developed: the most pressing things must be done first, and the others, though necessary, may have to wait a while. It makes a lot of sense to calculate, as best one can, Taguchi loss functions for the various candidate-processes, and then concentrate for the time being on those which have the steepest loss functions over their current normal operating ranges.

Fourth, again accepting that we are not talking in terms of any kinds of absolute truths, use of Taguchi loss functions does provide a possible basis, if needed, for putting figures on values of improvement efforts, and also for raising awareness of the costs of some current practices.

Two examples, briefly described here and then revisited in more detail in the next chapter, are a strategy for dealing with tool-wear and a problem with a cutting process. Both are cases where, because the processes initially adopted were based on the common judgment of quality in terms of meeting specifications, the real costs of the output were exceptionally high.

First, we take a process where tool-wear causes a gradual reduction in a quality measurement. A common practice is to set the machine to produce near the Upper Specification Limit when the tool is new, then let the mean gradually decline, and then replace the tool when the output approaches the Lower Specification Limit. The result is that overall output spans virtually the whole of the specification interval, with just as much at the edges of the interval as in the middle. The average Taguchi loss is therefore very high.

Second, there was the case of a company which was having trouble with the output from a cutting operation, in spite of the fact that the operation was conforming 100% to specifications. When measurements were taken of the lengths of the metal rods being cut in this operation, the remarkable observation was made that, although all measurements were indeed within specifications, by far the *majority* of them were either very close to the Upper Specification Limit or very close to the Lower Specification Limit, with hardly any of them near the nominal, halfway between those limits! In other words, the output was producing a vastly higher Taguchi loss than could ever have been imagined possible from 100% within-specifications material. The usual expectation is that most output is close to the *nominal*, with relatively few occurring in the extremes.

The explanation in this case turned out to be remarkably simple, and easy to act upon in a way which both increased the speed of the operation and reduced the average Taguchi loss several-fold. Clearly, a rod which is too long (above the Upper Specification Limit) is salvageable—cutting a bit off the end will bring it within specifications. On the other hand, a rod which is too short (below the Lower Specification Limit) is useless, at least for the current task. Consequently, the policy had always been adopted in this operation to set the mean length of cut near to the Upper Specification Limit rather than near the nominal. The rods were measured as they came off the production line; those below the Upper Specification Limit were immediately released for use, while those above the Upper Specification

Limit had a short length cut off them—that short length being pretty much the difference between the Lower and Upper Specification Limits. So those which were released immediately had lengths just below the Upper Specification Limit, while those subject to rework finished up just above the Lower Specification Limit!

As it turned out, the cutting process was in statistical control and eminently capable—i.e. its natural variability was considerably less than the difference between the specification limits. So the solution was simply to set the mean cut to nominal. The result was not only that no rework was subsequently needed, but virtually all the rods then produced were closer to nominal than to either specification limit. The reduction in average Taguchi loss was vast, and the troubles which users of the rods had been reporting ended overnight.

As indicated above, these examples will be examined in more detail in the next chapter, along with some more mathematical discussion of the use and implications of the Taguchi loss function. Non-mathematical readers may therefore regard the next chapter as optional.

For completeness, we should point out here that we sometimes encounter situations where loss functions are one-sided, as illustrated in Figure 35. This is the case in many service applications, where error-rates are important quality-characteristics. Other such examples involve percentage impurities in chemicals, and times which we would like to be as short as possible, e.g. down-times, loading-times, etc..[7]

Let us now make a few final comments on the use of specification limits. Some of Deming's thoughts reported earlier in this chapter recognise that specifications have had their uses in times gone by. But no longer:

---

[7] This paragraph was contributed by my BDA colleague Richard Kay, Director of Sheffield Statistical Services.

"Conformance to specifications:
a sure road to decline."

loss

measurement

FIGURE 35
A One-Sided Loss Function
(Richard Kay)

He also regards the matter seriously enough to include it amongst the Obstacles to the Transformation—see "The supposition that it is only necessary to meet specifications" in *Out of the Crisis*, pages 139–141, and near the end of Chapter 3 in this book. Incidentally, there is a strong link between that and the following Obstacle: "The fallacy of Zero Defects" (*Out of the Crisis*, pages 141–142). He also links these elsewhere:

"Zero Defects,
meeting specifications (incoming and outgoing),
are not good enough."

A particular drawback of dependence on specifications is that, unlike the use of the Taguchi loss function, specifications really provide no guidance at all to improvement efforts. The Taguchi loss function is an extremely useful approach to studying and quantifying goodness, accuracy, suitability of any quality-measurement of product or service; it is an approach which is appropriate in the new economic age. The Taguchi loss function received rather short measure in *Out of the Crisis*; Deming promises that it will attract much greater coverage in his next book.

Finally, a sad story indeed about the results of a company's dependence on, and belief in, specifications as the definition of quality (see also the story from a seminar delegate, reported on page 129 of *Out of the Crisis*, about the furniture company which tried to copy a Steinway piano!). The company decided to make copy machines. They carefully measured every component on a well-known Japanese make of copier, and laid down the specifications. They took a chance on the patents: they were prepared to pay the damages. There were 828 parts in all and, after a development time of two and a half years, and expenditure of some $36,000,000, every component was manufactured to those tight specifications. They were then put together, and:

"Everything fine—
except they wouldn't make copies.
Otherwise, they were all right!"

Chapter 12

THE TAGUCHI LOSS FUNCTION
—SOME DETAILED STUDY

The graph of the Taguchi loss function shown in Figure 34 is a parabola having a vertical axis and minimum value 0 occurring at the nominal value of the quality-characteristic. The equation of such a parabola is:

$$L(x) = c (x - x_0)^2$$

where x is the measured value of the quality-characteristic, x_0 is its nominal value, $L(x)$ is the Taguchi loss function, and c is a scaling factor (so that the loss can be expressed in whatever monetary units are deemed appropriate). It is the most natural and simple mathematical function suitable for representing the main features of the loss function as developed in Chapter 11.[1] Of course, that doesn't make it "right" for any particular application; note that, for example, it assumes the seriousness of departures from nominal follows the same pattern on both sides of the nominal (a particular exception to this assumption has already been discussed near the end of the previous chapter). Also, whereas the model is often reasonable over the specifica-

---

[1] Some statisticians may find the analogy with least squares techniques to be appealing.

tion range and for a moderate distance outside the specification limits, it is unlikely to hold at great distances from nominal: hopefully, however, our processes are not so bad that we need to be concerned about that.

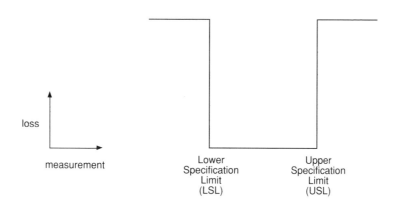

loss

measurement

Lower Specification Limit (LSL)

Upper Specification Limit (USL)

FIGURE 36
Loss Function Representing the Conformance-to-Specifications
Interpretation of Quality

But, even if the model isn't quite "right," it is surely bound to be closer to reality than a loss function representing the conformance-to-specifications approach, as seen in Figure 36. This shows supposedly zero loss throughout the specification range, with sudden jumps at both specification limits. Following the discussion in the previous chapter, there is no need to comment further on that, except as follows. Recall our observation in Chapter 11 on the perceived importance of specifications becoming a self-fulfilling prophecy. If there are systems, be they mechanical or bureaucratic, which "kick in" as soon as something goes beyond specifications, those very actions are costly. So there *is* then a sharp increase in loss as a specification limit is crossed —but that loss is generated by the management system itself, not by the quality-level of the actual product or service.

We shall use the parabolic model to study in more detail

some of the concepts and examples discussed in Chapter 11. Because it is only a model, the exact numbers produced in the calculations are not particularly significant. Minor differences will not mean anything important—a strategy which produces slightly greater loss than another strategy under this model could easily be preferable under a different model. But when we see differences of whole scales of magnitude—e.g. where the loss from one strategy is 10, 50, or even 100 times greater than that of another —then it is surely pretty safe to conclude that those differences are important, even if the parabolic model is just an idealisation.

As a further idealisation, which is necessary in order to produce the numerical comparisons in this chapter, we shall need to assume that processes can be *exactly* stable. Recall from Chapter 4 that the term "exactly stable" implies the process's statistical distribution is "steady, unwavering"; in particular, we can then talk in terms of *true values* of the process mean and standard deviation, which we shall denote (in this chapter only) by the traditional symbols μ and σ respectively. (This is, of course, in contradiction to Deming's important observation concerning real processes on page 334 of *Out of the Crisis*.)

We shall work in terms of the average (mean) Taguchi loss. The average Taguchi loss of a sample or batch of n items, whose values of the quality-characteristic x under study are: $x_1, x_2, \ldots, x_n$, is:

$$\frac{1}{n} \sum_{i=1}^{n} L(x_i) = \frac{c}{n} \sum_{i=1}^{n} (x_i - x_0)^2 .$$

If the process is exactly stable, with the quality-characteristic having probability density function $f(x)$, the average Taguchi loss can be computed as:

$$\int L(x) f(x) \, dx = c \int (x - x_0)^2 f(x) \, dx ,$$

which is the area under a graph of the product of the loss function $L(x)$ with the probability density function $f(x)$. Some straightforward mathematics converts this expression to:

$$c\left\{ \sigma^2 + (\mu - x_0)^2 \right\}$$

the terms in { . . . } being respectively the square of the standard deviation (usually referred to as the *variance*) and the square of the *bias*. It should be noted that, therefore, the average Taguchi loss does not depend in any complicated way on $f(x)$: it can be computed just from the simple parameters in this latter expression.[2]

To aid comparisons, let us also introduce the notion of the *capability* of a process. This is defined in different ways by different companies, but we shall take it here, in effect, as:

$$\frac{\text{difference between Upper and Lower Specification Limits}}{\text{difference between Upper and Lower Natural Process Limits}}$$

where for "Natural Process Limits" we can use the "true" 3σ limits for individual observations, so that the denominator can simply be written as 6σ. [3]

A capability of 1 (*unit* capability) is a bare minimum for the process to almost always be able to meet specifications.[4] A process is sometimes described as being *capable* or *incapable*

---

[2] An important consequence is that it is not necessary to make any assumptions about the form of $f(x)$, e.g. that it represents a normal (Gaussian) distribution. We have however used the normal distribution for illustration in Figures 37–40, and also for some fine detail of the computations and descriptions of the last two examples in this chapter.

[3] This is not Deming's definition of capability. Not suprisingly, he regards the capability of a (stable) process simply as being defined by the Natural Process Limits—without reference to specifications.

[4] For example, if the process is accurately centered, and the distribution is normal, then on average one measurement in just under 400 will lie outside specifications, usually by only a small amount.

according respectively to whether its capability exceeds 1 or is less than 1. Conventional thinking in the Western world is that a capability of 1⅓ is extremely satisfactory, and 1⅔ is perhaps extravagant since the probability of a measurement falling outside specifications is then negligible.[5]   However, note that the references to Japanese processes in Chapter 11 translate into capabilities of the order of 3 or 5. Also, for the measure of capability to reflect what the process may actually produce (rather than that of which it is capable), it has to be assumed that the process is accurately centered—i.e. that the process mean coincides with the nominal value x_0. We shall examine below what happens when this assumption is not satisfied.

We shall scale our results (i.e. choose the constant c in the equation for the parabola) so that a process of capability 1, accurately centered on the nominal, has average Taguchi loss of 100 units. First, let us look at values of the average Taguchi loss for exactly stable processes centered on x_0 but having various capabilities.

Table 1: Exactly Stable Process, Accurately Centered

Capability	½	¾	1	1⅓	1⅔	2	3	5
Average Taguchi Loss	400	178	100	56	36	25	11	4

We see that improving the capability to 1⅓ or 1⅔ does reduce the average Taguchi loss to roughly a half or a third of its value under unit capability. However, improving the capability to 3 or 5 brings *huge* reductions—describable in terms of "scales of magnitude," as we have referred to them earlier. Graphs of the average Taguchi loss against capability for all the examples to be studied in this chapter are shown in Figure 41.

---

[5] The currently-fashionable "Six-sigma" corresponds to a capability of 2.

The importance of accurately centering the process is quickly seen by comparing the figures in Table 1 with those in Table 2 below. The figures in Table 2 are derived assuming the process is (wrongly) centered on a point half-way between the nominal and one of the specification limits.

Table 2: Exactly Stable Process, Centered on Point Half-way Between Nominal and Either Specification Limit

Capability	½	¾	1	1⅓	1⅔	2	3	5
Average Taguchi Loss	625	403	325	281	261	250	236	229

Such bad centering totally ruins the potential advantages of improved capability. However, even with this bad centering, capabilities of 2 or more produce virtually no items outside specifications (so such processes would be regarded as eminently satisfactory under the conformance-to-specifications judgment), yet the assessments in terms of the Taguchi loss function are massively worse than if the process is accurately centered; for example, when the capability is 2, the loss in Table 2 is *ten times* that in Table 1.

We shall now consider the two examples described at the end of the previous chapter. First, we had the tool-wear problem. Let's recall the details. The process is initially set to produce output near the Upper Specification Limit (USL). Tool-wear then gradually reduces the measurements, and the process is stopped and the tool replaced when the values get close to the Lower Specification Limit (LSL). Note that the capability (except for the drift problem) of such a process must be greater than 1 in order for this scheme to exist, for otherwise there would be no room for such manoeuvring. However, for completeness, we have included the case of unit capability in the results below.

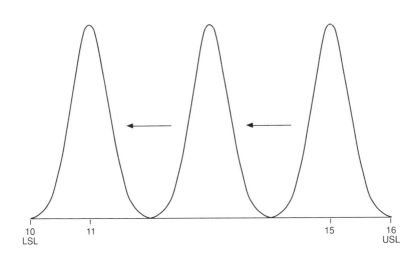

FIGURE 37
Process Drift: Capability = 3

Figure 37 demonstrates the case where the capability is 3. For illustration, we are taking the values of the LSL and USL to be 10 and 16 respectively, and the standard deviation σ to be ⅓ (instead of 1, which would have been the value of σ if the capability were 1). We shall initially center the distribution at 15, so that the distribution is just below the USL. Suppose the process mean then drifts steadily down to 11, at which time we stop the process, replace the tool, and reset at 15 again. (If the capability of the process were 2 instead of 3, i.e. $\sigma = 0.5$, we would have to initially center the process on 14.5 and allow it to drift down to 11.5 before replacement, as shown in Figure 38.) The average Taguchi losses for processes with various capabilities which are "controlled" in this way are shown in Table 3A. (Cost of replacement of the tool is not explicitly included in the computations.)

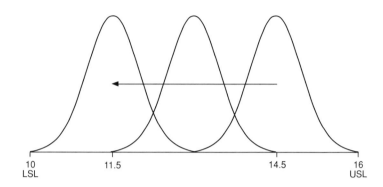

FIGURE 38
Process Drift: Capability = 2

Table 3A: Process with Uniform Drift,
Started and Terminated in Order to Just Avoid
Out-of-Specifications Measurements

Capability	1	1⅓	1⅔	2	3	5
Average Taguchi Loss	100	75	84	100	144	196

What a surprise! For low values of the capability, the average Taguchi loss does decrease, but then it soon starts to *increase*, so that there is almost twice as much loss when the capability is 5 as when the capability is 1! Upon reflection, the reason for the increase becomes clear. When the capability is high, the process begins very close to the USL—and is thus bound to produce items well away from nominal, which therefore causes high Taguchi loss. The same is true with respect to the LSL just before the tool is replaced. Because of the quadratic form of the loss function, the harm caused in these extremes overrides the benefit of the good product produced half-way through the operation.

Note that this result is in direct contrast to the view obtained in the conformance-to-specifications world. The scheme is organised so that essentially no out-of-specifications material is produced, whatever the capability (as long as the capability exceeds 1). Increased capability has the advantage that the process runs for longer before needing to replace the tool; however, as we now see, that "saving" is false economy in terms of Taguchi loss. The average Taguchi loss could be substantially reduced by, say, replacing the tool twice as often. As an example, in the case of a capability of 3, this would allow the process to be initially set at 14 (rather than 15), and to have the tool replaced when the process mean has dropped to 12 (rather than 11). The average Taguchi loss would now be 44 instead of 144—still, of course, nowhere near as good as a process of capability 3 without drift (for which the average Taguchi loss is 11; see Table 1) but a very considerable improvement on what happens when we wait as long as possible before replacing the tool. Table 3B shows the results of halving the tool-replacement times for the same capability values as used in Table 3A.

Table 3B: Process with Uniform Drift,
Replacing Tool Twice as Frequently as in Table 3A,
and Centering Process as Far as Possible

Capability	1	1⅓	1⅔	2	3	5
Average Taguchi Loss	100	61	48	44	44	52

Are the substantial improvements in average Taguchi loss compared with Table 3A worth the cost of changing the tool twice as often? That's for management to judge.

Finally, we come to the cutting operation. Recall that the process mean was set higher than nominal because of the undeniable logic that it is easier to make a long rod shorter than to make a short one longer! Let's model this case by assuming the mean cut is set at the USL and that, if a rod's length is found to exceed the USL, then an amount equal to the specification range (i.e. the difference between the LSL and the USL) is cut off. Yet again, this is a very simplified model, but the results are interesting and do relate pretty well to the practical case which stimulated this investigation.

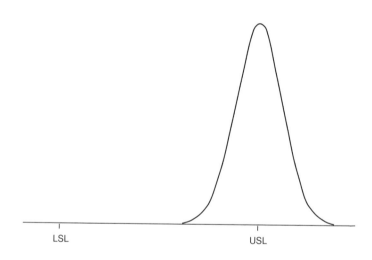

FIGURE 39
Cutting Operation: Distribution of Original Cut

The problem with the scheme is easily indicated by a couple of pictures. The initial cut has a distribution such as that shown in Figure 39. After the second cut is made from those rods (half of them) which are too long, the lengths of the resulting rods have the distribution shown in Figure 40!

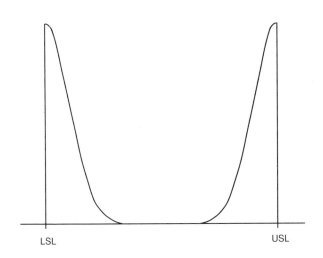

FIGURE 40
Cutting Operation: Distribution After Rework

Table 4: Cutting Operation, Centered on USL
with Rods Above USL Reworked
(by Cutting Off an Amount of (USL–LSL))

Capability	½	¾	1	1⅓	1⅔	2	3	5
Average Taguchi Loss	343	439	521	597	649	686	752	808

It is immediately obvious why the average Taguchi loss is so high—see Table 4. Most of the lengths are close to the specification limits, and few if any are near the nominal. In other words, most of the lengths yield close to the maximum Taguchi loss possible for within-specifications material, while there are hardly any near the zero-loss mark. It should also be obvious to the reader that this is another case where increased capability actually makes the results worse.

As we can see, this system (which made quite good sense from the viewpoint of conformance to specifications) produces absolutely appalling results in terms of the Taguchi loss function.

As stated earlier, Figure 41 shows graphs of the average Taguchi loss for all the examples which we have studied in this chapter. Enormous differences are clearly seen—differences which are almost entirely hidden if we are content with conformance to specifications.

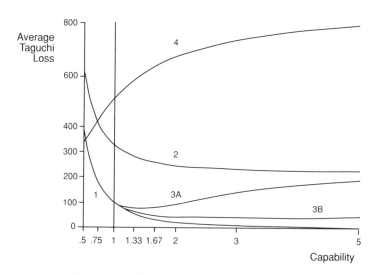

(The curves are labelled with the number of the corresponding table.)

FIGURE 41
Graphs of Average Taguchi Loss

PART 3

THE NEW CLIMATE

THE NEW CLIMATE

As we said in Chapter 3, Dr. Deming's teachings are not cast in stone. He is responsible for a living, dynamic, and continually-developing philosophy of management which is based on foundations that are solid, sensible, and true.

Out of the Crisis was published in 1986, around the time of Deming's 86th birthday. Most people don't continue much learning or development of new thinking after their 86th birthday. But Deming, as I know well from my meetings with him during recent years, *is* still learning, and expecting to learn; and he is still developing the thoughts which have their origins back in the 1920s when he first became aware of Shewhart's fundamental work on the nature of variation. At one recent seminar I heard him say:

"I have learned more in the last three months than in any other three months of my life."

Deming is a fine example indeed of practising what he preaches in terms of the first of his 14 Points, which calls for constancy of purpose for ever-continuing improvement. Compared with him, I fear that most of us have our thought-processes pretty much fossilised by the time we are half his age.

To some people, even Deming's ideas of a few years ago seemed too radical and ambitious. However, the modern world has seen much change: some good, some bad. As pointed out in the

Preface to this book, quite a lot of the change of the past ten years or so has been in line with his teaching. And, by and large, it has been change for the *worse* which has been *out of* line with his philosophy. So, when we now examine Deming's recent developments and new concepts, which are likely to seem outrageous or downright impossible to the sceptics, let's at least keep an open mind and preferably a positively receptive one. He has been right so many times before. And, certainly, if the world can change in the direction he wants, it will be a much happier place in which to spend our declining years than that which most other development theories will produce. But:

> *"Nothing will happen without change."*

And we're all on the hook! Addressing all who hear what he has to say:

> *"Your job is to manage the change*
> *necessary to create the new climate."*

This Part of the book describes the new climate.

Chapter 13

JOY IN WORK

Joy in work. Where does *that* appear in BS5750 (ISO9000), Juran, or Crosby?[1] Where indeed did it appear in Deming prior to 1988? It was seen only indirectly through the tamer language of "pride of workmanship," which has been the subject of the 12th of the 14 Points since they were formulated at the start of the 1980s. And maybe even that hasn't received the attention it deserves, through being overshadowed in many people's perception by the reference in the same Point to the abolition of performance appraisal.

Yet, during 1988 and beyond, this is where Deming has started many of his presentations:

"Why are we here?
We are here to come alive,
to have fun, to have joy in work."

As an example, near the beginning of a seminar in London in July 1988, Deming said:

---

[1] Philip B. Crosby is another American who is well-known as an orator on quality. He is, for example, identified with the concepts of Cost of Quality and Zero Defects which however, it may be noted, have no place in Deming's work.

*"The aim of management, management's job,
is to enable everybody to enjoy his work."*

And, in Denver a month later:

*"Management's overall aim
should be to create a system in which everybody
may take joy in his work."*

Likewise, during the first two minutes of the British television documentary *Doctor's Orders* (previously referenced in a footnote in Chapter 9), we heard:

*"Management's job is to create an environment
where everybody may take joy in his work."*

Now, of course, the work in the four-day seminars is study, learning, education. Therefore we need fun in learning — all learning:

*"A school should be like a laboratory
—with excitement in learning."*

Somewhat darkly, Deming points out that another way to have fun is for someone to ask a stupid question, as that will provide fun for all of us; volunteers are needed. Those who have attended one of his four-day seminars will know what he means!

Who but Deming would have thought, or dared, to raise such a far-reaching, outlandish, unrealistic concept as "joy in work"? And what a world of meaning and implication is contained within those three little words (rather similarly to Point 8's "Drive Out Fear"). Yet, having raised the controversial notion that this *is* the "job of management," it seems to me that here we have a vital missing link in Deming's previous teachings. How can we achieve "Constancy of Purpose" for continual improvement (Point 1) without joy in work? How can we "Adopt the New Philosophy" (Point 2) without joy in work? Recall the words of Noël Coward, playing the role of Captain "D" in the film *In Which We Serve*:

> "A ship can't be happy unless she's efficient, and she certainly can't be efficient unless she's happy."

We shall concentrate in the next chapter on the essential role of innovation in the survival and future success of a company. Compare the chances, if you will, of important innovation being made by those who have, and those who do not have, joy in their work.

In Chapter 29 we shall look in detail at an extensive list of "Faulty Practices" of management which Deming abhors— such as performance appraisals, and the use of MBO and arbitrary numerical targets. These are, in effect, deliberate introductions of conflict, competition, and fear—the direct opposite to the Cooperation: Win-Win culture which is the subject of Chapter 15.

Why are these practices so common, and indeed approved of by many in management? The answer is that they are all examples of *making the best of a bad job* (see Chapter 3): when the management and working environment is bad, such practices do (at least, on the surface) make things less bad than they were before. The concept of "joy in work" is irrelevant, even ridiculous, in this context. But Deming is not concerned with the short-sighted task of making the best of a bad job—he is concerned with the far-sighted objective of transforming the bad job into a good one, a very good one. And joy in work plays a large part indeed in that context.

Consider motivation. We hear a lot about "motivation of the work-force" these days. There are all sorts of programmes ostensibly concerned with just this, including EI (Employee

Involvement), EPG (Employee Participation Groups), and QWL (Quality of Work Life). A common incredulous reaction from those encountering the 14 Points for the first time is: how on earth can we motivate people if appraisals, fear, targets, incentives, threats, and exhortations are to be removed? An abrupt answer from Deming is that if management stopped *de*motivating their employees then they wouldn't have to worry so much about motivating them.

Suppose management are successful in what Deming now calls their job—to enable, encourage, and engender joy in work; what need will they then have to concern themselves with motivation? It will already be there. Create joy in work, and EI will come; get joy in work, and QWL will follow—we won't need programmes. If people have joy in their work, they are fuelled by intrinsic rather than extrinsic motivation (see Part D, Items 5 and 6 in the System of Profound Knowledge in Chapter 18); they become first-class citizens, responsible to themselves; and they become able and willing, indeed enthusiastic, to contribute to the "Four Prongs of Quality" which will be introduced in the next chapter. Conversely, those suffering in a system which does not enable and encourage them to have joy in their work will not, dare not, so contribute. And joy in work has largely been smothered by modern management and the political and social environment in which we live. No wonder we are on the decline. For:

> *"The prime requirement*
> *for achievement of any aim,*
> *including quality,*
> *is joy in work."*

Joy in work, as indeed is the case with many of the constituents of the Deming philosophy, is not a new idea: it is more a forgotten idea. But it comes on good authority. My great friend and colleague in the British Deming Association, Professor David

Kerridge of the University of Aberdeen, has pointed out these extracts from the book of Ecclesiastes:

> "There is no happiness for man but to eat and drink and to be content with his work. This, too, I see as something from God's hand." (Chapter 2, verse 24)

> "When man eats and drinks and finds happiness in his work, this is a gift from God." (Chapter 3, verse 13)

> "I see there is no happiness for man but to be happy in his work, for this is the lot assigned him." (Chapter 3, verse 22)

(These versions are taken from *The Jerusalem Bible*.)

How is it that some people seem to think it is somehow immoral to enjoy work? And that, for work to be valid, it has to be *suffered*? And that if people seem to be happy in their work then some kind of crackdown must be called for?

Deming suggests that maybe 2% of management and 10% of shop-floor personnel have joy in their work; I fear he is being overly optimistic. Can *you* think of anybody whom you could genuinely describe as having joy in his work? Maybe you will draw a blank. However, if you can find a few candidates, just think of the invaluable contributions that those people make to the companies or other organisations for which they work. They are beyond price. Just think then how things would be improved if Deming's stated percentages were correct. Think even further on what things would be like if a quarter or a half of both management and other employees could be described in that vein. It would be "out of this world." Yes, it would—it would be transformation. But transformation, no less, is precisely what Deming is talking about: "the transformation that is required for survival" is referred to even on the flyleaf of *Out of the Crisis*,[2] let alone in the text itself.

As a final thought, why *should* people do a good job instead of merely time-serving and getting away with the minimum they can? I'd suggest three possible reasons:

1. fear;
2. financial incentive; or
3. they want to.

Which do you think will be the most effective?

---

[2] This refers to the American edition; regrettably, most of the flyleaf text is omitted from the Cambridge University Press version.

Chapter 14

INNOVATION—
NOT JUST IMPROVEMENT

Some approaches to quality are versed in terms of reaching, and then retaining, some standard of product which is supposed to be satisfactory to the customer—meeting specifications or other types of requirements (see Chapter 3). To do that "at minimum cost" (actually a contradiction in terms) is naturally regarded as a better economical aim. Maybe these ideas can be extended to services as well as products. In the old days, the approach by which such standards and lowest cost were supposed to be achieved was unclear, but one could attempt to protect the customer by mass inspection (an immediate complicating issue for calculations on minimising cost). It is even more unclear how such concepts could be extended into service.

Fundamental in Deming's and some others' teachings of old was the change of emphasis from sorting finished product by inspection, i.e. "downstream" action on output, to "upstream" action on processes, i.e. to improvement of efficiency and quality so that end-inspection is less necessary. This also made it more feasible to apply quality considerations to service as well as to products. However, most people's interpretation was to improve processes just enough to meet required standards and specifications. The overriding feeling was still that, by and large, process and quality improvement imply extra cost rather than cost-savings. It comes as a shock to most Western managers who read the early pages of *Out of the Crisis* to find that Deming was telling the Japanese as early as 1950 that improved quality leads to *lower* cost and *increased* productivity.

So there is a huge leap to make—away from the notion that quality is all about meeting specifications and other requirements and standards, and toward the philosophy of ever-continuing improvement. This is emphasised in the first and fifth of Deming's 14 Points, which together call for constancy of purpose for continual improvement of quality in products and services and in all processes which feed into them—in short, in all operations and activities within the company. What does continual improvement of quality mean in this context? Some express it as continual reduction of variability. Certainly, that is a vital part of the story, though it is not all.

So that's it: continual improvement—*that's* what Deming is all about. And many people believe that. But wait, even in *Out of the Crisis*, let alone in his more recent work, it becomes clear that continual improvement is *not* all that Deming is talking about. A hint comes as early as page 11 of his book:

> "Statistical control opened the way to engineering
> *innovation.*"

(my italics). However, the heading of that section is still worded: "Innovation to *improve* the process." But on page 25, in the discussion on the first of the 14 Points (Constancy of Purpose), comes the clear, unambiguous message:

> " ... constancy of purpose means acceptance of obli-
> gations like the following: ... Innovate ... Plans
> for the future call for consideration of:

> New service and new product that may
> help people to live better materially,
> and that will have a market;
>
> New materials that will be required; ...
>
> One requirement for innovation is faith that
> there will be a future. Innovation, the foundation
> of the future, can not thrive unless the top man-
> agement have declared their unshakable com-
> mitment to quality and productivity. Until this
> policy can be enthroned as an institution, middle
> management and everyone else in the company
> will be sceptical about the effectiveness of their
> best efforts."

And, on page 135:

> "It is thus not sufficient to improve processes.
> There must also be constant improvement of de-
> sign of product and service, along with introduc-
> tion of new product and service and new technolo-
> gy. All this is management's responsibility."

So the message was there in *Out of the Crisis*; but, in recent years, it has become much louder and stronger. Following the above quote from page 135, Deming now frequently refers to the "Four Prongs of Quality" as:

1. Innovation in product and service;
2. Innovation in process;
3. Improvement of existing product and service; and
4. Improvement of existing process.

Even in those companies which have some sense that they need "to do something about quality," there is usually very incomplete comprehension of the above list. A common error is to think that it can all be done by Number 4—i.e. just by improvement of processes. Everything working at top speed, without problems, faultless—then we're a quality company. If only it were so easy. It would all be up to the workers! (Not so, of course, but more to them than with the other three Prongs.)

Think of just about any product or service, and suppose that you are now providing it as perfectly as could be imagined— except that its design, features, and facilities are five years out-of-date. It may well have taken you five years to advance the fourth Prong to that stage of excellence. Great—unfortunately, you are likely to be out of business.

Maybe there are exceptions. Maybe that particular product or service is still in fashion, and there is still a great market for it, but for some reason nobody else is providing it any more. In that case, you had better jump in and make your fortune!

No: good processes and good operations are essential, but they are unimportant in the sense that they are not the decisive ingredients for quality in the eyes of the customer. Improvement in the design of the product or service is also essential. And innovation in process, product, and service is both essential and important. The product or service must have a market. Prongs 1, 2, and 3 help to ensure that, but not Prong 4.

The opportunity for innovation, and the environment in which it flourishes, does not just magically appear. It has to be developed and nurtured by management. This is one of the main reasons that Deming frequently says:

"Quality is made in the Board Room."

So the Four Prongs are all essential. They are listed in order of importance; but the order of application has, in practice, generally to be the reverse. It is dangerous to try to invent new processes when our current processes are behaving badly and we don't understand why. It is foolhardy to produce new ranges of products or services when our current products and services are unreliable. For, if what we now offer to our customers is lousy, what confidence can they (or we) have that our new products and services will be any good? After all, we surely know more about what we are using and producing now than we do about potential innovations. So, if we're doing a poor job now with what we know, what sort of job would we do with what we don't know?

Incidentally, what clearer example of this message could we have than that of the Japanese? The complaint for a long time was that all the Japanese did was to steal our ideas. What is there in a Japanese car which was not first made in the Western world?—

"The Japanese simply did it better."

—small if needed, more snappy design if needed, more fuel-efficient if needed, more powerful if needed, and more trust-worthy, as reliability surveys in consumer magazines have been showing continuously for the last decade. Yes, the Japanese were willing to learn, and *how* they learned! Is there supposed to be something wrong with that? Somehow, many people seem to feel that there is. Learning is education and, for reasons which I do not claim to understand, a lot of people in Britain seem to regard education with some disdain. Sadly, many educators themselves seem to have a pretty limited view of education, including the need for their own continuing education.

So the Japanese just learned, rather than creating their own innovations? Well, whether that was true or not, who can doubt that they are innovating now,[1] and will do so increasingly in the future? Think of optical goods and sound technology, for

---

[1] Sometime after completing this chapter, I noticed the newspaper headline: "Japanese drive off with top awards in car trade 'Oscars' " (*Daily Mail*, London, 23 November 1989). The article commented: "Japan took half the prizes in the car world's first 'Oscars' ceremony yesterday, eclipsing its rivals for innovation, initiative, design ... Honda ... won the award for innovation with a 'two-in-one' engine that cuts out half the cylinders when they're not required ... "

examples. It's no use complaining about the past—we are where we are. We need to innovate, and to make the best of our innovations. Incidentally, note how far we've moved from thinking of quality merely in terms of the meeting of specifications!

But, even if the Japanese are now into innovation, they still can't come up to us, can they? No? For a round figure, Deming warns that we only have perhaps 10% of the innovation which we need.

Where does innovation come from? Where do the ideas for innovation come from? From the customer? By asking him what he wants? No. The customer often doesn't know what he needs *now*. He certainly doesn't have much idea of what he's going to need in the years to come. If *we*, as producers of product or service, wait until then to find out, clearly we are always going to be several steps behind him, and pretty big steps at that—and behind our competition who are more forward-looking. Unlike in past times:

"Necessity is not the mother of invention."

The world moves faster than that now. As we saw in Chapter 3, for future survival, let alone success, we need to *stay ahead* of the customer. Innovation does not come from the customer:

"How would he know?"

Let's look at things from the customer's point of view. If we consider any particular innovation, however essential it may be to the way we work and live our lives now, the fact remains that both we and the world in general were doing fine prior to that innovation. People cannot miss what they do not know about. Ignorance, so the saying goes, is bliss. Maybe we cannot conceive now of existing without the fax machine, the cordless telephone, the word-processor, convenience food, the microwave oven, supermarkets, felt-tip pens, synthetic fibre, synthetic *any-*

thing—but we used to. We coped—that was the way of life with which we were familiar. We, the customers, did not imagine that such advances were possible—they didn't even cross our minds. What important innovations are you currently standing in line for? Any? Probably not. But, when (good) innovation does become available, from whatever source, then the customer discovers that he needs it. If it is only available from one source, that's where he has to go to get it. If it is available from two or more sources, the customer will obviously start looking for the best value. Even more obviously, any source from which it is not available is not in the running.

When innovation appears, it comes from the producer who looks ahead to find answers to the question: what new product or service will help the customer, will be attractive to the customer, will entice the customer, will whet his appetite? If a producer does not produce innovation, he may no longer be a producer. Management need to create a climate where innovation can flourish. How?

There are many links to more familiar features of the Deming philosophy in the answer to that question, and also to the other newer emphases in this Part of the book. First, hear his clear message about how *not* to do it. One of the many faults of performance appraisal is its encouragement to "live by the book," to "stay in line," and fit in with whatever the boss would like. The effect must be to stifle initiative and progress:

"People get rewarded for conforming.
No wonder we are on the decline."

(see also Chapters 30 and 32). The sequence of the argument is that the "do as you're told" atmosphere of performance appraisal, especially if it is coupled with the need to push others under in order to get a good rating, destroys any chance of joy in work; for joy in work needs, in particular, a sense of security and confidence, and these are the very prerequisites for an environment in which innovation can thrive. I am reminded of Deming's answer to a question at a seminar in Nottingham during 1988: "Does your philosophy apply to R&D organisations?" It was:

> *"Any good research has to be done*
> *by people who have joy in their work."*

Particularly for real innovative research and big break-throughs (as opposed to more routine "mapped-out" research), Deming believes that a climate of both individual and company security and joy in work is essential. He claims that all 80 American Nobel Prize winners have been in positions of effectively total security, rather than having their attention continually diverted by short-term pressures (of which managers who are still in the old way of thinking would often approve in order to "keep people on their toes").

One of those Nobel Prize winners (for Physics in 1956) was William Shockley who, jointly with J. Bardeen and W. H. Brattain, developed the transistor in 1948. Other specific examples he mentions of breakthrough innovation from positions of total security are the invention of nylon by Du Pont and ICI working together ("Win-Win" again), Faraday's work on magnetism in London around 1820, Mr. J. C. Capt's work on the Census, and Irving Langmuir who, working in the General Electric laboratories, started development of the carbon filament electric light (and he wasn't even looking for it); Deming also mentions Shewhart's fundamental work on variation in this same respect. Then there was Harvey Firestone:

> *"People thought he had lost his mind:*
> *riding on air—how silly!"*

Deming says he remembers the first such tires from around 1910. Of course, they weren't too good—one had to carry around liquid cement or some such substance for repairs—but what a breakthrough; what innovation!

So this is Deming's answer to how the climate for innovation can be developed. Concentrate on generating joy in work, and the power of intrinsic motivation will lead to innovation, thus enabling the company to grow and prosper. As so often the case: so few words, but such a depth of meaning. Deming says that North America has done a "pretty good job of smothering innovation"; it has all but lost it:

> *"We have smothered innovation by*
> *patchwork, appraisal,*
> *putting people into slots,*
> *rugged individualism."*

The rest of the Western world should learn, and not reproduce the damage.

What companies are in the best position to improve quality and to open wide the gates of innovation? Not the ones who are turning to quality as the last resort for survival, i.e. those who have seen the writing on the wall, and may be too late. For those companies, through that very situation, are restricted, constrained, held back by their problems of today. Sure, danger sharpens the appreciation of the need for change, but the breathing-space and flexibility conducive to genuine long-term change for the better are just not there in the company whose days seem to be numbered. Companies that are doing well, the future assured as far as anyone can see, have that breathing-space and flexibility to improve and innovate, and thus to further help their own economic condition and that of those around them. Not only do they have the greater opportunity:

> *"This kind of company*
> *has the greatest obligation to improve."*

An inevitable logical conclusion from this call for positions of security is (opposite to most current opinion) the desirability of monopolies or "fortresses of power." This is one of the several issues on which shallow thinking can result in impetuous rejection of Deming's message. But, as we know, Deming's thinking is anything but shallow, and any critique of it needs to be based on much deeper consideration. Development toward monopolies, like abolition of performance appraisals, and of fear, and of arbitrary targets, and of mass inspection, are natural *consequences* of the development of Deming's principles, and will be good only in the appropriate environment; in the wrong environment, they can obviously cause more harm than good.[2]

For example, eradication of mass inspection, the aim of Point 3, would be crazy whilst processes, systems, quality of raw materials, relationships with suppliers, etc. are bad. But mass inspection becomes redundant when processes have been brought into control, relations with suppliers have moved much nearer to partnership than conflict, and an environment of continual improvement has become established.

Similarly, "fortress of power" situations are only proper and generally beneficial when service and responsibility to the customer have become predominant, and a company's responsibility to the community which it serves and to society in general is well-established.

In all these cases, it is highly dangerous to put the cart before the horse (especially on an uphill slope!).

---

[2] Eve Williamson, in a letter to me on the content of this chapter, writes in terms of "the difference between massive, ill-managed bureaucracies and those organisations which know how to command great resources with autonomy of participating companies and individual and organisational learning of high quality." Let us hope that the top managements of the increasingly-powerful multinational companies and of the European conglomerates being formed in preparation for the 1992 single market have a true sense of their positions of high privilege *and* high responsibility.

Chapter 15

COOPERATION: WIN-WIN
BACKBONE OF THE NEW PHILOSOPHY

*"Everybody wins on better quality;
everybody loses on poor quality."*

Throughout the 1980s, Deming has been saying that what is needed is nothing short of "transformation of Western management." We have recently (in Chapter 13) seen his reference to transformation even on the flyleaf of *Out of the Crisis*. And his Preface commences:

> "The aim of this book is transformation of the style of American management. Transformation of American style of management is not a job of reconstruction, nor is it revision. It requires a whole new structure, from foundation upward."[1]

But what *is* the required transformation? It has been

---

[1] As in Chapters 1 and 3, we may surely broaden the geographical range of these statements. We should remember that, at the time of completing *Out of the Crisis*, Deming's work was still relatively little-known in the Western world outside America. For example, his first four-day seminar in London, in June 1985, attracted less than 100 delegates, only one-third of whom were British. Things are different now.

expressed in many ways. At the end of his discussion of Point 2 in *Out of the Crisis*, Deming writes:

> "Transformation is required—adoption of the 14 Points, and riddance of the Deadly Diseases and Obstacles."

This can be extended in an obvious way to find that the transformation needed is the adoption of the Deming philosophy! But in recent times Deming has given a more specific indication. It has become clear to him that there is a backbone to the required transformation, a backbone which supports the whole body. That backbone is the conversion from the old economics based upon conflict and competition (Win-Lose: I win, you lose, or you win, I lose) to a new economics based on cooperation (Win-Win: everybody wins). It is conversion from the mistaken belief that competition is inherently good for everyone—companies, their employees, and their customers—to the realisation that working together for mutual benefit and for the benefit of society at large has far greater potential. It is conversion from a current society in which both the cause and the effect of there being winners is that there must be losers to a future society in which there need be no losers and neither are any desired. Alfie Kohn's book: *No Contest* makes great reading on this subject. See also the beginning of *Out of the Crisis*, Chapter 4.

This is total transformation. Two terms which Deming likes to use are *change of state* (attributed to Ed Baker of Ford Motor Company) and *metamorphosis* (attributed to Barbara Lawton of Albany International)—change as vital and as irreversible as the emergence of a butterfly from a caterpillar or a fly from a maggot. The analogy can be carried further: without such change, the creature dies an early death!

What will such transformation do? It will open the door to optimisation rather than mere suboptimisation, making it feasible to realise the full potential of the system within which we work and live. It will make possible the very changes whose necessity we have discussed in recent pages. How can one have joy in a work-situation where success is only achievable by down-grading or destroying others? Joy in work can only come if your internal suppliers are doing their best for you, raising the circumstances and potential of your job ever-higher and thus enabling you to do your best for your internal customers. And joy in work both encourages and enables you to contribute improvement and innovation to your work and to that of those around you. Remember Deming's estimate that only two out of 100 managers take joy in their work (Chapter 13):

"Most of the 98 have their eye on a good rating, not daring to contribute innovation to their work."

At a seminar in London, Deming defined the very purpose of the 14 Points as being "to help carry out cooperation, to create joy in work."

Why has competition rather than cooperation become regarded as a good thing? Has cooperation even been tried? Oh, there are thousands, probably millions, of *examples* of cooperation, and we shall see some below. But, quoting ahead from the next chapter:

"No number of examples establishes a theory."

In the absence of a *system* of cooperation, i.e. of the adoption of cooperation as a theory, a principle, and an objective, a system of competition will be appealing and will be seen as effective. Of course it will—there is nothing with which to compare it. Also, a system of competition is easy to create and easy to manage. Governments can pass anti-monopoly legislation, and they can insist that local authorities accept lowest tenders, managements can install performance-appraisal mechanisms and bonuses and incentive plans, they can install man-of-the-month schemes,

they can install MBO and cover their company with arbitrary
numerical targets, and most teachers, lecturers,[2] and education
authorities use grading and ranking to place children and older
students into slots without knowledge of, or care for, the effects—
or of the messages from the Experiment on Red Beads (Chapter 6),
or of the unfairness of the Pygmalion Effect (Chapter 30).   In
management, it is often easier to develop strategy whose direct
aim is to choke a competitor rather than to create improvements
for the benefit of customers and of society as a whole.

In the American context, Deming sees reasons for the pre-
dominance of this Win-Lose philosophy in the country's historical development:

> *"Anyone that could run fast enough*
> *toward the West could find and preempt land—*
> *free, his—at the expense of someone else*
> *that did not run fast enough."*

He refers to this as "rugged individualism"; it is the sacred cow
of competition: "looking after Number 1."   As long as there was
still free land available, such a philosophy might not have been
all bad; he who lost might live to fight another day, and there
was still some booty as yet unclaimed.   In other countries, different analogies may be found suitable.   Similarly, management by a
philosophy of conflict and competition is not all bad in an ever-
expanding market.

But such a philosophy is not all good either.   And when
the free land runs out, and when the market doesn't carry on automatically expanding, then the bad surely outweighs the good.   In
that situation, Win-Lose implies much destruction: by definition,
one company's or person's win is at the expense of another's loss.
Actually, the destruction is much greater because the struggle to

---

[2] i.e. professors (in the context of North America)

win in the short term (for survival) uses up resources which could otherwise be employed for long-term advancement.

The current apparent need for competition, rather than being an automatic fact of life, is an indication of the scale of transformation needed. Of course, competition provides protection, of a kind, for the consumer. As is the case with all the "Faulty Practices" of management (see Chapter 29), it exists and is perceived as good because, as suggested earlier, it makes the best of a bad job. The real task is instead to transform the bad job into a good one so that the bad practices become redundant (again like mass inspection after appropriate improvements have become established).

Think of the resources and energy which are wasted in competition, think how much effort is duplicated, think how often wheels are reinvented. Think instead of the rewards which would accrue if that energy could be expended in cooperation rather than conflict, motivated by joy in work and desire to serve society rather than just trying to beat the other fellow.

Now I doubt whether Deming, however forward-looking he may be, conceives of a world without competition. Yes, we are faced with competition, on a national or global scale, and are likely to remain so as far as we can see. But our aim cannot be merely to meet the competition, else we shall always be behind. Our competition does not stand still. How can we reach ahead? *Not* by creating yet more competition internally (within the company, or within the country, or on whatever scale we are thinking) with all its consequent waste, but instead by more genuine cooperation aimed at having everybody win.

Does this imply that everybody's advantage will be the same? No, there are bound to be inequalities: some will win less than others, but that is better than their losing. Never mind so much about merely increasing market share, i.e. grabbing a bigger slice of the pie—let's cook a bigger pie: let's expand the market. In so many areas, foreign competition is thinking this way. If we don't, and if we are not hungry for it, they'll cook that bigger pie *and* take all the extra for themselves.

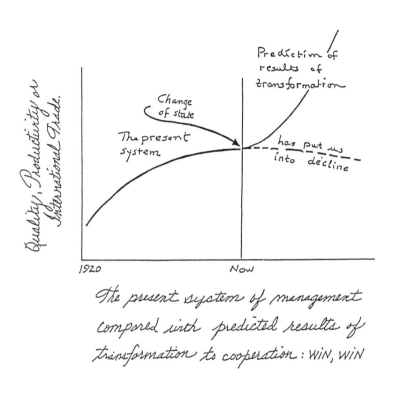

FIGURE 42
The Present System of Management Compared with Predicted
Results of Transformation to Cooperation: Win-Win
(Drawn by Dr. Deming)

Competition may be considered to have served us well enough over recent decades—but then, many things have served us well enough in their time: e.g. Taylorism, conformance to specifications, bloodletting, carrier pigeons, and the abacus. So with competition. In his 91 years, Deming has seen more things come, serve their time, and go (though not quite all of those just mentioned!) than most readers of this book; believe me, he has a pretty keen eye on what needs to go now. His thinking, covering the period from his own youth to now and into the future, is compactly summarised in his drawing of Figure 42.

Actually, Deming is not so charitable about competition as I have just been; indeed, he makes no bones about it:

"Competition has ruined us."

If we concentrate on crushing our competition (and I am thinking of our *local* competition—what chance do you think we have of destroying foreign competition?), what is the overall gain or loss? Are those who are thrown out onto the streets to be left hungry? Of course not; but who pays? We may seem to be winning for a while—in some relative, local context. But will it last? What chance is there against foreign competition whose inclination, within their own country, is toward the Win-Win philosophy?

Deming's thesis is for everybody to win; that there be no losers. A vital part of the transformation is the realisation that competition—getting ahead by pushing others under—can be ruinous to us all. So, instead of the easy management system of competition (and it *is* easy—why shouldn't they take the lazy way out if they have no theory to guide them?), people in management need to learn and to practise cooperation as a *system*, as a *principle*, as a *strategy*, as an *objective*:

"One of the first steps in the transformation will be to learn about cooperation: why, what, and how."

Cooperation needs to become a major influence in policy-making. It shouldn't be incidental or an accident: it should be the *aim*. This needs to become the new management style. It is transformation—from rugged individualism to real cooperation: Win-Win, no loser. It is nothing less.

Before proceeding, we should clarify what is *not* meant by cooperation here. The word is sometimes used to mean mere patience, forbearance. For example, if our flight is delayed and we've been held up in the gate-area for hours, we may eventually be thanked for our cooperation. But:

> *"That's not cooperation.*
> *What can prisoners do?"*

The word is also sometimes used with evil connotations— for collusion, for dividing up markets to the *dis*advantage of customers. This is not what we are talking about.

The theme of genuine cooperation is compactly summarised in one of the vertices of the Joiner Triangle: "All One Team" (see Chapter 3). This is a foundation-stone of the Deming philosophy. The spirit of "All One Team" is magnificently reflected in a quotation from a production worker (*Out of the Crisis*, bottom of page 79) who clearly reflects his feeling of *belonging* to the company—he is *part* of that company. It is with sadness that, during industrial disputes, I hear broadcasts of union representatives talking of "the company" and meaning the *management* of the company, the workers being on the other side of the fence. Do read that quotation from *Out of the Crisis*, and then reflect on Deming's observation that:

> *"People on a job are often handicapped*
> *by lack of privilege*
> *to work together with people*
> *in the preceding and following operations."*

This is what lack of teamwork means in practice. The new philosophy requires leadership of people (Chapter 25), a vital part of which is the nurturing of teamwork.

Yet again we become aware of the barrier that merit-rating forms to real progress. Indeed, Deming has referred to the merit system itself as a "manifestation of the Win-Lose philosophy," and has charged that system as being one of the main constraints holding us back from the Cooperation: Win-Win culture. The worst case is performance appraisal constrained by a forced distribution, where it becomes formally necessary to put somebody else under in order to get a higher rating. A forced distribution creates artificial scarcity: there can be only so many of high rank.

Dr. Deming has pointed out to me the analogy with a game such as tennis: when somebody wins, somebody else loses—that's the rule of the game. But it creates artificial scarcity of winning, *no matter how excellent the players.* Does life in general, and work in particular, have to be played like a game of tennis? Surely a better way is to have everybody working for the company rather than against each other or against other departments in the company.

Deming has recently further described the needed transformation as "a new system of reward." Here I would like to quote a paragraph from his paper which was read in Osaka on 24 July 1989 and which summarises so much of what I am now calling the backbone of his philosophy:

> "The transformation will be a new system of reward. The aim will be to unleash the power of human resource contained in intrinsic motivation. In place of competition for high rating, high grades, to be Number 1, there will be cooperation between people, divisions, companies, governments, countries. The result will in time be greater innovation, science, applied science, tech-

nology, expanded market, greater service, greater
material reward for everyone. There will be joy
in work, joy in learning. Anyone that enjoys his
work is a pleasure to work with. Everyone will
win; no loser."

Departments in a company are usually managed in a system of competition. (Ponder the fact that the words "departments" and "divisions" are often used analogously!) Yet every
department is a supplier or a customer, or both, to every other
department. In an inner-competitive environment, what matters
the trouble which you cause to the other departments? Indeed, in
such a situation it can actually be *advantageous* for you to cause
trouble to other departments (or at least as much as you can get
away with).

"All One Team" also needs to be extended outside the
company: preferably, as far as possible, into customers and certainly into suppliers. More generally, as the industrial world
expands, considerations of efficiency and economic production
move us inexorably to greater and greater interdependence. Is it
better for that interdependence to be based on conflict and distrust, or upon cooperation and partnership? As implied in
previous chapters, "partnership" is the word which Deming has
started using to describe the desired state of customer-supplier
relations (see Chapter 22): "arms around" not "at arms length"!
And "arms around" tends to imply a sole supplier, or a sole customer, or both.

Deming quotes many instructive examples of cooperation,
ranging in scale from localised to worldwide. On the world scale,
he cites international conventions in such varied areas as the
date, the time of day, red and green traffic lights, AA and other
regular sizes of batteries, the metric system, and focal lengths of
lenses. As he points out, the world was not *made* this way. These
are not coincidences; the wind does not blow voluntary standards
in our direction. These and other huge conveniences, which reduce

costs and make life better and easier for everybody, are there because of human action and agreement, national and international cooperation, and the endeavours of standardising bodies, working together with manufacturers and consumers.

Who could doubt that, in spite of the disintegration of the USSR, worldwide cooperation has a role to play in avoiding nuclear war? And should we not have some sense of cooperation with future generations as regards conservation of resources and the prevention of environmental catastrophe?

Examples of more localised international cooperation are seen in the formation and development of the European Community, in standards for cross-border railway gauges, and in mutual transfer of information on methodology used in national censuses (an area in which Deming has been involved since 1939). An example of cooperation in the European Community is the Meteorological Office at Bracknell in England, working with pooled resources of all member-countries and serving each country with a far better forecasting service than any one of them could manage to achieve individually.

Note that, when international bodies and industries fail to develop a needed standard, the only alternative to chaos is a government regulation. That is much less satisfactory than a voluntary standard. Voluntary standards are more likely to be constructed by people who have knowledge which can be used for general benefit and convenience; people involved in the construction of government regulations are less likely to satisfy these criteria. Yet voluntary standards are voluntary, and government regulations are compulsory.

Deming cites examples in the American car industry concerning safety, fuel economy, and catalytic convertors. The industry itself should have worked together on such matters, to provide greater economy and effectiveness for their customers. They did not. So the Government produces its regulations, and each motor company scrambles individually to meet them. The costs and losses are unknown and unknowable. Let us learn. To survive, there is much on which we need to agree nationally and internationally: better voluntarily than forcibly.

Suppose that managements of airlines would work together with themselves, their customers, and the relevant aviation authorities regarding the use of common facilities at airports. Instead, most airlines have their own equipment at airports, thus incurring expensive excess capacity to cope with peak loads. Of course, there is some cooperation. Sometimes one airline engages another to handle baggage or carry out repairs. Or one company checks in the passengers for many airlines. This is good, and there should be more of it; it would both increase profit and give better service: Win-Win. I fly into Heathrow Airport, London quite often, and had for a long time been impressed at the speed with which baggage was unloaded and delivered (whether or not flying British Airways) compared with what I suffer at other airports. It was Deming who pointed out to me recently that British Airways handles not only their own baggage but the baggage of many of the airlines at Heathrow.

There are a number of other stories which Deming is fond of telling. One comes from page 32 of W. G. Ouchi's *The M-Form Society*. Ouchi was attending the annual three-day meeting in Florida of an American trade association. Having observed how the meeting's sessions tended to adjourn at around noon for golf, fishing, etc., he pointed out that the association's Japanese counterpart was currently meeting for some 60 hours per week, and had been doing so for over three months, to agree on standards, including safety standards, so that they could approach their Government with one voice in order to expedite appropriate legislation, export policy, and finance. As Ouchi said to his audience in Florida: "Tell me who you think is going to be ahead five years from now: you or your competitors?" I have heard some people refer to such cooperation as unfair competition!—

"The Japanese are born into a life of cooperation. Is North America to be left behind?"

Even for those who cannot conceive of the transformation from Competition: Win-Lose to Cooperation: Win-Win on anything like the scale that Deming sees it, it still makes so much sense to cooperate on common problems:

"Compete? Yes, but in the framework of cooperation first. Everybody wins."

Such cooperation, like the development of international and other standards referred to above, doesn't necessarily abolish competition. If competition is still the order of the day, initial cooperation for the sake of improved conventions, standards, and quality still makes things better for society at large.

A second example recalls an occasion in 1960 when General Motors suffered serious fire damage at their Hydra-Matic plant near Ann Arbor, Michigan. Within 24 hours their competitors, Ford and Chrysler, had put on three shifts a day in order to bail them out, and sent technicians to aid reconstruction at the GM plant. Naturally, GM paid for this help, but the result was that they lost not a single car through the incident.

The mention of General Motors reminds me of the fine cooperation between the United Auto Workers and the management in General Motors' Quality Network. Of course, this did not instantly materialise: much work, study, discussion, and planning was involved. I recall attending a ceremony on 20 July 1987 at the BOC (Buick-Oldsmobile-Cadillac) Powertrain headquarters in Brighton, Michigan: it was for the dedication of their impressive new Dr. W. Edwards Deming Conference Center. I began talking with the person I sat next to, and was soon impressed at his deep knowledge and understanding of Deming's work. It was only later that I discovered he was a UAW official.

In a third story, Deming gets very close to home. He recounts how his car refused to start one day. He phoned his local garage to arrange for the car to be towed away for repair. The garageman duly arrived driving *his competitor's* pick-up truck. It turned out that the two competitors had just one pick-up truck each, and shared them whenever the need arose. The result was that customers received virtually the same service as if the two garages *both* maintained *two* pick-up trucks. However, by such cooperation, the savings for both garages were considerable, to the benefit of both businesses and their customers.

A related example is that of two service-stations, on the same corner, which Deming observed took turns to stay open late at night. Obviously, customers were attracted to that corner outside normal business hours, and then more generally. Another example of Win-Win: customers were happy, and the owners of both service-stations were happy, since they were both making greater profits than would have been the case without the arrangement.

Shopping at the The Dalles, a town along the Columbia river west of Portland, Deming observed that, in any store which did not have what he was looking for, the sales-assistant would recommend a competitor who would probably be able to satisfy him. He does not know whether there was a general agreement amongst the storekeepers, or whether one just started the practice and the others followed suit. Again, everybody wins—storekeepers and customers alike. I believe that this practice is quite common in America. Regrettably, in my experience, it appears much less common in England.

At a four-day seminar in Denver, Tom Boardman of Colorado State University expanded on the theme of fruitful cooperation between competing companies. Specific areas of cooperation for mutual benefit which he discussed were:

> on increasing the markets;
> on improving the industry image;
> on using interchangeable parts;
> on agreeing on ANSI standards;
> on working with common vendors; and
> on improving the process for producing products
> and services.

Let us recall Deming's example in Chapter 8 of the railway company that expanded its service by the addition of a motor-freight business. We saw that optimisation of the expanded system would be different from separate optimisations: that was the point of the merger. Could that improvement of service have been achieved if the companies had stayed separate? For:

> ***"Optimisation (Win-Win) would be the same,***
> ***whether they be independent***
> ***or joined in ownership."***

Thus, *in theory*, there should be no difference in operation whether or not the two businesses were under the same management. But would the best be done by separate companies? Could it be done? The two managements would need to agree on aims and methods just as if they were one company. And, if they tried, wouldn't there be interference from the Anti-Trust Division, on suspicion of price-fixing? ("Anti-Trust"—now that's an interesting choice of word!)

The sad fact is that, going by the way many companies are run, progress toward optimisation would not occur even if the businesses were joined. Knowledge and desire for the benefits of such optimisation, along with true cooperation, are essential. And:

"Accomplishment of this aim requires use of a System of Profound Knowledge."

(see Chapter 18). However, the chance of progress is likely to be higher under combined management than as separate companies. Similarly to that which we saw in Chapter 14, the logic draws us to the necessity for mergers in order to optimise service and profit. On the other hand, complexity is bound to increase with the size and diversity of the organisation. Is there a limit, and what is it?—

"The answer lies, I believe, in the ability of the human mind to understand the benefits of cooperation, and knowledge to manage optimisation."

A further possibly-startling result of the same logic concerns the very price-fixing just mentioned. Quoting from Deming's communication of 3 January 1990 (referred to in Chapter 8):

> "If management were wise, and cared enough about service and profit, and possessed the requisite knowledge for overall optimisation of service and profit, price-fixing would bring maximum benefit to everybody including, of course, customers. Everybody would win. The Anti-Trust Division would be superfluous. Does the Anti-Trust Division teach optimisation, everybody win?"

Let me hasten to repeat (see the end of Chapter 14) that all these thoughts are *consequences,* and rather long-term ones, of the adoption of the Deming philosophy; clearly, they cannot stand alone.

Sadly, the belief has grown, nurtured by politicians, that business is not about cooperation but rather about competition and conflict: competition between companies, competition between departments in the company, competition between individuals, conflict with suppliers. Even in soap-operas, it is common to find friends or family-members "doing each other down" with the apparently respectable explanation: "That's business."

We all know the reasons why competition is accepted as good for the consumer. But is it necessarily so?—

"America has been sold down the river on competition."

Deming cites the undesirable consequences of the American Government's deregulation of the airlines and telephone and land-freight services: poorer quality and higher prices. Prices may be lower for a while—that's competition. The lower prices are paid for by lower quality. Unfortunately, the quality stays low but the prices do not. Cutting internal costs cuts internal quality, and that eventually raises costs. Who pays? Guess! In the meantime, companies that had more concern for quality may have been squeezed out of business by the price-war. The consumer may well be sorry about that, but too late. Lose-Lose.

What role *should* Government play in business? For a start, those who judge Deming as taking a position which is too far left in the political spectrum should have their fears dispelled by the unambiguous statement:

"Certainly we don't wish for any government agency to run a business."

What do government agencies do? What should they do? Some exist to take care of failures—necessary, but immediate admission of failure of the systems which governments, in large part, administer:

> *"Certainly we need to take care*
> *of failures of business, including banks,*
> *and of individuals,*
> *the hungry, the homeless."*

But which government agencies take on responsibility to help management of companies to cooperate, leading to better business —essential for future success, even survival? Surely governments should, for example, encourage cooperation on common problems,[3] such as disposal of waste and other necessary services. That would not be government *in* business:

> *"It would be government in favour of business,*
> *not against business."*

Referring again to the deregulation of air fares by the Civil Aeronautics Board and of freight rates by the Interstate Commerce Commission, Deming points out that in both cases deregulation just suddenly became law, throwing business into chaos. He submits that it would have been much better to give the airline and freight industries time (in the airlines' case he suggests four years) to *prepare* for deregulation, trying to bring the companies together to work on common problems for better profits and *better service to the public*: Win-Win. For the freight

---

[3] Another pertinent observation from Eve Williamson is that, in Britain, the Training Agency has done more than most government departments to foster cooperation and collaboration through a number of initiatives, to persuade industry "to get its act together," and to recognise and understand the necessity for genuine investment in people. Unfortunately, as the name implies, the present emphasis is only on training: there seem to be no complementary initiatives to nourish education (as distinct from training, in the sense in which we understand these words in this book).

business, wouldn't it have been better to get the management of the railways and road-freight together, to plan for their mutual benefit and that of their customers? Instead, they are in competition as always.

In Britain, we know well the effects of such competition. Over a quarter of a century ago, Dr. (later Lord) Beeching put into effect a plan for making the railways "slim and fit." What do we see now? An inadequate road system clogged with heavy goods vehicles, and a press-report on my desk as I write, forecasting further substantial rises in British Rail fares following a 21% increase at the beginning of this year.

Deming notes that the American Government harasses rather than helps businesses, and in particular is highly obstructive to their cooperation. We know why, of course—to prevent the possibility of monopoly, automatically regarded as bad for the consumer. But need it be (see Chapter 14)? As a case in point, he often tells of the Pacific Automotive Cooperation, headed by an ex-Chairman of Toyota, formed to help both manufacturers and consumers get better parts. The PAC was set up in Toronto. Why not in Chicago, or Atlanta, or Detroit? The answer is clear. Because of the Anti-Trust legislation, such an activity would be illegal in America. Deming talks of other examples where he knows of CEOs of companies, wishing to cooperate for their mutual benefit *and for that of their customers*, having to conduct their business on the golf-course because their aims would also be regarded by the authorities as illegal or, at least, highly questionable.

Yet again, it is clear why the legislation is there. In Britain, there is similarly the Monopolies and Mergers Commission. An environment and culture has been developed in which even our largest and most respected companies are not to be trusted. There must be a better way.

EXAMPLE:
HUGE FINANCIAL ADVANTAGES OF COOPERATION

Harm that is caused by internal competition and conflict and the fear that is thereby generated, and good that is brought about by internal cooperation and teamwork, is of massive proportions. A Purchasing Manager, under pressure to reduce his figures, changes to a cheaper source, even if he buys poorer products and service as the result. Engineering Design imposes unnecessarily tight tolerances to compensate for the fact that Manufacturing never reaches the standards asked of it. Departments performing better than budget start spending near the end of the year because they know that otherwise their next year's budget will be reduced. As the end of the month looms, salesmen start doing everything they can to meet their quotas, with scant regard to the problems caused to Manufacturing, Administration, and Delivery, let alone to the customer. Figures get massaged, computations "redefined," so that reports show more of what senior management want to see. Some (not Deming) might summarise this in terms of the CYA syndrome holding sway. (CYA was recently interpreted for us by Myron Tribus as "Cover Your Anatomy"!) Yet again, suboptimisation destroys any chance of optimisation.

The following simple illustration clearly indicates something of what is respectively won and lost in environments of cooperation and conflict. It is based on an example that Deming has used since 1987, on which I have elaborated with his approval.[4]

---

[4] The original thoughts on this illustration came from Mr. Fred Z. Herr, Vice-President of Product Assurance in Ford Motor Company, in conversation with Dr. Deming in July 1988. I am grateful to Nida Backaitis,

For brevity and ease, the illustration is on a small scale. It is not difficult to imagine how the numbers multiply in more realistic examples.

	Effects of Options			
Areas and their Options	Effect on Area A	Effect on Area B	Effect on Area C	
Area A				
Area B				
Area C				

FIGURE 43

Example: Huge Financial Advantages of Cooperation —Table 1

The blank chart shows that the illustration concerns an

of the University of Southern California, for her help while I was enlarging upon those original ideas during a four-day seminar in Denver during August 1988.

organisation comprised of just three areas or departments. In the left-hand column are listed options available for each area to adopt or not adopt, according to their choice. The remaining columns show the potential effects of adoption of the options.

Areas and their Options	Effects of Options		
	Effect on Area A	Effect on Area B	Effect on Area C
Area A			
i	+		
ii	+		
iii	+		
Area B			
i		+	
ii		+	
Area C			
i			+
ii			+
iii			+

FIGURE 44

Example: Huge Financial Advantages of Cooperation —Table 2

When there are barriers between the areas (hierarchical management style, MBO, inner competition, etc.), each area naturally adopts options which are beneficial to itself.

However, let us examine the effect of this primitive decision-making mechanism both on the individual areas and hence (by summation) on the whole company. For ease of illustra-

	Effects of Options			
Areas and their Options	Effect on Area A	Effect on Area B	Effect on Area C	Net Effect on the Company

Area A
  i
  ii
  iii
	+
	+
	+			

Area B
  i
  ii
		+
		+		

Area C
  i
  ii
  iii
			+
			+
			+	

Net Effect of Adopted Options				

FIGURE 45

Example: Huge Financial Advantages of Cooperation —Table 3

tion, let us assume that each + or – corresponds to gain or loss of the same amount of money throughout.

Areas and their Options	Effects of Options			
	Effect on Area A	Effect on Area B	Effect on Area C	Net Effect on the Company
Area A				
i	+	−	−	−
ii	+	−	+	+
iii	+	−	−	−
Area B				
i	−	+	−	−
ii	+	+	−	+
Area C				
i	+	+	+	+++
ii	−	−	+	−
iii	−	−	+	−
Net Effect of Adopted Options	++	− −	0	0

FIGURE 46

Example: Huge Financial Advantages of Cooperation —Table 4

Options which are locally beneficial to one area may well be quite the opposite for other areas. Suppose each area's gains and losses are as indicated. The net effect on the company happens to be zero in this case, i.e. equal to the effect of doing nothing at all (which would be a rather easier way of achieving the same result!). Depending on the details, the net effect could have been positive, zero, or negative.

	Effects of Options			
Areas and their Options	Effect on Area A	Effect on Area B	Effect on Area C	**Net Effect on the Company**
Area A				
i	+	–	–	–
ii	+	–	+	+
iii	+	–	–	–
Area B				
i	–	+	–	–
ii	+	+	–	+
Area C				
i	+	+	+	+++
ii	–	–	+	–
iii	–	–	+	–
Net Effect of Adopted Options	+++	+	+	+++++

FIGURE 47

Example: Huge Financial Advantages of Cooperation —Table 5

If the management environment improves so that areas become aware of the effects of their actions on other areas (they may have been aware already, of course, but it wasn't in their interests to pay any attention) *and* the old inter-departmental rivalries are replaced by genuine teamwork for mutual and company benefit, options are now only adopted if they produce *net benefit to the company*. Consequently, only three of the previous eight options are now adopted. (Entries relevant to adopted options are characterised in the Figures by bold print; the options now rejected have been shaded over.) By this more judicious and restricted choice of actions (i.e. wise choices of both action and inaction), the company is much better off and, in this new management environment, *everybody* gains.

Areas and their Options	Effects of Options			Net Effect on the Company
	Effect on Area A	Effect on Area B	Effect on Area C	
Area A				
i	+	−	−	−
ii	+	−	+	+
iii	+	−	−	−
iv	−	+	+	+
v	−	+	+	+
vi	−	−	+	−
Area B				
i	−	+	−	−
ii	+	+	−	+
iii	+	−	+	+
iv	+	−	+	+
Area C				
i	+	+	+	+++
ii	−	−	+	−
iii	−	−	+	−
iv	+	+	−	+
v	+	−	−	−
Net Effect of Adopted Options	++++	++	++++	+++++ +++++

FIGURE 48

Example: Huge Financial Advantages of Cooperation —Table 6

Further, in the improved environment, options which previously never saw the light of day are now considered. These are options which are locally *dis*advantageous to the area which can adopt them but which give benefit to other areas. Amongst this greater range of options, again the ones to be adopted or not adopted are now chosen according respectively to whether they are or are not of net benefit to the whole company. The bottom line results speak for themselves.

* * *

It can be an illuminating training exercise to get people, after they have seen this example, to identify options that they or their departments have or have not adopted recently, and to discuss the +/– implications in other parts of the company. Of course, the exercise will be more fruitful in companies where the environment is already improving; in less good contexts, people are often unwilling to spill the beans, even in the classroom. Old habits die hard.

Deming is of the opinion that the situation sketched in Table 2 (Figure 44) describes most of the world. He mentions a company which is divided up into four companies, each supplying all the others. Each is rated on its own profits! Such foolish practice is common. North America does not understand the above exercise:

"Cooperation is not the American way."

He said in a London seminar that he believes it is more the British way. I hope he will be proved right, despite all the pressures to the contrary. Experimental evidence from Carlisle and Parker (*Beyond Negotiation*, pages 46–51) unfortunately indicates the opposite. Their repeated use of the "Prisoners' Dilemma" or "Red-Blue" game in both America and Britain indicates *less* willingness on the part of the British than the Americans to cooperate for the general good.

Deming summarises the lessons to be learned from the above exercise as follows: [5]

---

[5] This is an extract from a privately-circulated paper: "Foundation for

"Options for various companies to consider, or for the managers of the various divisions within a company, must be stated, and enumerated. Approximate effects of action on each option stated, yes, no, or leave it alone, for every staff area involved, must be computed.

A new set of options might turn out to be better for everybody. Ingenuity is required to generate the options to consider. This is management's job.

Divisions should not operate without recognition of their interdependence. They will otherwise almost with certainty make decisions that are far from optimum for the company as a whole, and everybody will lose."

And recall from Chapter 8 that:

"It would be poor management, for example, to purchase materials at lowest price, or to maximise sales, or to minimise cost of manufacture or design of product, or of service, or to minimise cost of incoming supplies, to the exclusion of the effect on other stages of production and sales. All these would be suboptimisation, causing loss. All these activities should be coordinated to optimise the whole system."

"All these activities should be coordinated to optimise the whole system." That's what Cooperation: Win-Win is all about.

Management of Quality," dated 17 September 1988.

PART 4

FOUNDATIONS OF KNOWLEDGE

FOUNDATIONS OF KNOWLEDGE

As we continue our study of the Deming philosophy, let us pause to consider what "philosophy" means. Chambers Twentieth Century Dictionary describes "philosophy" as "pursuit of wisdom and knowledge" and as "knowledge of the causes and laws of all things." *Knowledge* is the link. And knowledge is the focus of this Part of the book.

I also like two of the descriptions of "philosophy" from the less-fashionable *Readers Digest* Universal Dictionary: "the investigation of causes and laws *underlying reality*" (my italics) and "enquiry into the nature of things based on logical reasoning rather than empirical methods." So again we see an implicit warning against depending too much on empirical data. That final definition also leads us pretty directly into Deming's affirmation in Chapter 16 that both experience and examples teach us *nothing* unless they are studied with the aid of theory.

Experience and examples are no replacement for theory; worse than that, they may well mislead. We recall the harm caused by tampering—best efforts which look good to the unwary but which make things worse. Chapter 17 returns to the familiar theme of "best efforts," along with all the other "obvious" ideas we saw in the Author's Foreword on how to improve quality.

And finally, we come to a summary and discussion of what Dr. Deming calls a "System of Profound Knowledge." Deming sometimes surveys this System early in his four-day seminar, describing it as a "bird's eye view" or an "architectural drawing" of what is to follow. In this book, the System of Profound Knowledge serves as a pivot between the variety of topics discussed so far, both those current at the time of *Out of the Crisis* and those which have been developed since, and a return visit to the 14 Points with which we shall complete the book:

> *"The 14 Points for Management*
> *in industry, government, and education*
> *follow naturally as application of*
> *the System of Profound Knowledge,*
> *for transformation from*
> *the present style of Western management*
> *to one of optimisation."*

Chapter 16

THEORY AND THE SEARCH FOR
EXAMPLES

Deming often describes the purpose of his four-day semi-nar as *education* (see Chapter 31); and the purpose of education is to learn *theory*. Maybe a few come to the seminar looking for something easier; something more straightforward—a recipe or a formula. They are soon set straight as regards that! For:

"This may be a new experience:
you have come to learn.
You may have come for a formula.
There is no formula.
There is no Step 1, Step 2, Step 3, ...
We are going to learn a whole lot more.
We're going to learn theory.
We're going to learn why we have to do
what we need to do."

Before long, delegates hear of the "new philosophy" (Point 2), and leadership (Point 7), and teamwork:

> *"The new philosophy requires leadership.*
> *Leadership will nurture teamwork.*
> *We shall try to understand the theory*
> *behind these statements."*

And the reader will recall from Chapter 1 the requirement to start out with confession: not of sin, but of stupidity:

> *"Stupidity?*
> *Only theory can help."*

Now nobody, including Dr. Deming, is claiming that theory alone is sufficient to discover the reasons for problems and difficulties, and to identify important opportunities for improvement:

> *"You don't find out*
> *without knowledge and theory.*
> *And you don't find out if you're not there."*

He often makes a big point of asking his audience how all the horrible examples (including five from Japan) reported in *Out of the Crisis* got into the book. The answer that he awaits is: "Because you were there."

Insistence on a foundation of theory is one of the things that marks Deming out from other supposed experts on quality. His teachings are not just a collection of good ideas, or challenging ones, or controversial ones (though they certainly are all of these). They are rather outcomes and logical deductions based on theory. When that is realised, his work is seen to have a unique unity and credibility. Deming points out how grateful he is that he worked for those ten early years in the laboratories of the United States Department of Agriculture, giving him the opportunity to "seek new knowledge and test new theories."

We should say a little about what is meant by the word

"theory." To some it is perhaps merely the opposite of "practice," and it may thus be given little credit or attention—rather analogously to Ian Graham's description of common practice as a tiny "Plan" and a huge "Do" rather than a proper PDSA Cycle (see Chapter 9 and also the comments about the "practical man" in Chapter 31.)

Let us look at our dictionaries again. They are not much help: we find as definitions of "theory" everything from "speculation as opposed to practice" to "beliefs or suppositions serving as a basis for action"! It is the latter end of this spectrum which is appropriate to Deming's use of the word; his use is intimately tied in with practice. One should also beware of confusion with the mathematical use of "theory," implying valid and logical deductions (which is relevant to our use) from axioms which need have no relevance to the real world (which isn't).

The "beliefs or suppositions" mentioned above may range from mere hunches to well-researched and well-established hypotheses. Actually, we should not talk of "mere" hunches, since surely some of the greatest breakthroughs began just as hunches sometime:

"A theory may be complex. It may be simple.
It may only be a hunch,
and the hunch may be wrong.
We learn by acceptance,
or by modification of our theory,
or even by abandoning it and starting over."

It is not hard to see a pretty clear hint of the PDSA Cycle in that statement! Any theory is better than none. Formulating a hypothesis, and then comparing it against practice, is fundamental to the Scientific Approach.

A great advantage of using theory is that one does not have to judge separately the consequences of theory "on their own merits." Recall the train of thought already developed in Chapter 4: if the theory is unarguable and our processes of deduction valid, the consequences must also be unarguable. One might take, for example, the Joiner Triangle as a theory. (Note that Deming himself does not do this, but the reader will know from Chapter 3 that I find it helpful.) *If* you believe that the three vertices of that Triangle are correct foundations on which we should build for the future, and *if* you see that much of the Deming philosophy, including in particular the 14 Points, naturally follows from those foundations, how can you argue against the philosophy?

Deming notes that a great value of theory comes from the way it sometimes enables us to predict results without experimentation, saving time and money. And, to revisit a familiar theme:

"Any psychologist could have predicted the result of introducing the merit system."

Models used by statisticians and others constitute theories. According to George Box, however: "No model is correct ... ," at least in our real world, " ... but some models are more useful than others." Similarly, Deming points out that any theory is correct in its own context, i.e. in its own world. But as regards this world:

"Is the theory helpful to us, especially for prediction?"

And there lies the crux of the matter: prediction. Deming's whole object is to improve the future. As far as he is concerned, *that* is the purpose of knowledge. The past is gone. We can learn from it, of course—with the aid of theory. But, be careful. Any statistician studying regression analysis knows that it is always possible to develop a model which *exactly* fits any (finite) amount of data from the past. He also knows, or should know, that that model is almost always totally useless as a predictor for the future. Thus, the value of experience is more limited than commonly thought. Even if circumstances remain the same into the future, the regression analogy just mentioned should warn us against relying too heavily on experience.

There is another crucial lesson to learn from regression theory. And that is the danger of extrapolation: if the circumstances change, the value of past experience may not only be "limited": such experience may well be dangerously misleading. This relates to Deming's teaching on enumerative and analytic studies in statistics (see Part B, Item 7 in the System of Profound Knowledge in Chapter 18). Think of Euclidian geometry. It works perfectly, as far as we can tell—so long as the distances are not too great. It *would* be perfect in a different world: a world where, in particular, the earth is flat, and lines and points have zero thickness. But it is not recommended that ships or aircraft in this world use Euclidian geometry!

Thus, Deming gives us frequent warnings about using experience without theory; in fact, to be blunt about it:

"Experience teaches nothing
unless studied with the aid of theory."

In Chapter 9 we followed this with: "Experience teaches us (enables us to plan, to predict) only when we use it to modify and understand theory." So, again, PDSA is seen as a central theme. A final nail in the coffin of experience alone is the observation that, if it *could* ensure success, the United States would be way ahead economically. America has far more man-years of experience in modern management of industry, both product and service, than any other country.

And, as with experience, so with examples. The "Search for examples" comes high in the list of "Obstacles" (see Chapter

3). Deming says the same thing about examples as he does about experience:

> **"An example teaches nothing**
> **unless studied with the aid of theory."**

And we are warned that:

> **"No number of examples establishes a theory."**

For those examples may all be in a context which does not relate to the future. One can think of a million examples consistent with Euclidian geometry: but the million-and-first might be out in space. And:

> **"A single unexplained failure of a theory requires**
> **modification or abandonment of that theory."**

The big danger with a search for examples is that most people do not seek to study them with theory, but seek instead to copy them without thought—or, at least, without sufficient thought. (Remember the example of mere copying mentioned in the final paragraph of Chapter 11.) Think of the countless excursions to Japan, and the number of times "the Japanese secret" has been brought back:

> **"People go to Japan, and learn nothing,**
> **for they have no theory to learn with."**

Closer to home:

> **"To copy without knowledge of theory**
> **a company that is doing well**
> **is to invite destruction."**

Deming is clear that some of the most successful companies "are successful by accident" while providing demonstrations of some of the worst practices of management. Of course, he has no objection to our studying successful, or unsuccessful, companies in order to plan and to learn to manage better—but that is not achieved by copying, only by studying with the aid of theory. We shall mention in Chapter 30 a company, "written up in books as highly successful" says Deming, which uses surely the worst merit-rating system of all: a forced distribution over as few as five people.

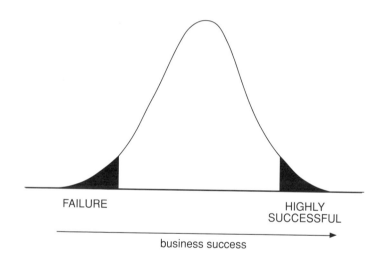

FIGURE 49
Distribution of Companies' Levels of Success
(Dr. Deming)

To emphasise the point, Deming sometimes draws a picture of a distribution representing the output of a stable system—see Figure 49. The horizontal axis records, in some sense, a company's level of failure or success. Companies in the right-hand tail are popularly praised as highly successful; those at the left are branded as failures. But this could as well be a distribution of red beads: high numbers on one side and low numbers on the other. Is it fruitful to copy "methods" for obtaining few red beads and avoiding those for obtaining many? An analogy which I have heard Deming use here is that of putting on the same clothes as a successful man. As he says, "It will take a little more than that":

> *"Only theory can help us figure out*
> *what's right and what's wrong."*

And with this theory of management:

> *"Need any country be poor?"*

What did Japan have, other than theory and people?

Chapter 17

QUALITY AND BEST EFFORTS

Let us now return to where we began in the Author's Foreword—to the quality referendum, and all the suggested answers as to how quality may be achieved.

We may now understand why all those answers were wrong:

"Every one of them
ducks the responsibility of management.
They require only skills,
not minds."

None of them requires knowledge, leadership, or theory of management. People delude themselves that they have knowledge, and such delusion is one of the biggest obstacles to progress; quoting Daniel J. Boorstin in *The Discoverers*:

> "The greatest obstacle to discovering the shape of
> the earth, the continents, and the oceans was not
> ignorance but the illusion of knowledge."[1]

---

[1] I am most grateful to Ian Graham for making me aware of this striking quotation.

The first four answers: automation, computers, gadgets, and new machinery, will tend to keep doing the same things as before, only faster: there will be greater productivity of the same problems, the same faults. Indeed, it is hardly uncommon experience to find that new machinery brings new problems with it. Deming mentions a company (unnamed, but many readers will guess correctly) which advertised that the future belongs to him that invests in it, and went ahead and spent tens of billions of dollars on new machinery. Most of it, says Deming, turned out to be a binge into high costs and low quality. The intentions were good (see below): the management obviously believed that a better future could be achieved this way. The President of the company wished Deming to go and see one of his factories with all its nice new automation. He was so sorry that Deming didn't have the time:

"I asked him whether he thought
that that would keep them in business.
He had never thought about it.
He was enthusiastically doing the wrong thing."

Some readers, referring back to the list of wrong answers, may have the idea that Deming is disrespectful of computers. Not so:

"Computers are wonderful—
if they are used right.
But more computers are not the answer."

He refers to a group of consultants who advertised:

"Computerised quality information systems provide the vital link between high technology and efficient decision-making."

With a tinge of regret in his voice, he remarks:

"I wish that management for quality were so simple."

The next three items on the list: making everybody accountable, and posters, slogans and exhortations, and Zero Defects, fall foul of the arguments concerned in the tenth of the 14 Points (Chapter 28)—in particular, they are based upon ignorance of the fact that the vast majority of problems are due to the system within which people work, and not to the people themselves (see Chapter 4).

The following four items: MBO, incentives and bonuses, Management by Results, and work standards and quotas, are the subject of Point 11 (Chapter 29), while the merit system and pay for performance are involved with Point 12 (Chapter 30).

Remember from the Foreword that some of the answers are wrong in the sense that they are insufficient. Others are wrong in a much stronger sense—they have a *negative* effect on quality: they stand in the way of progress. The ones just mentioned fall into that category. Further, Just-in-Time (the Kanban system) is fine if processes are in statistical control, but is dangerous otherwise; and the problem of (merely) meeting specifications has been fully discussed in Chapters 11 and 12:

"It's amazing how many are turning out product at maximum cost— helping to put us out of business."

I include the item "common sense" in the list as the result of Deming's response to a question at the meeting of the Association Française Edwards Deming in Versailles on 6 July 1989. A questioner asked: "What differences are there between the good common sense approach and the real quality approach?" Deming mused that "common sense" might be the most dangerous thing in the world. It has no meaning. Surely, common sense has led to all the "Faulty Practices" of management (see Chapter 29):

> *"They are our ruination.*
> *So, if all the Faulty Practices have come from*
> *common sense,*
> *we must beware of common sense."*

And that brings us to the final three items in the list: good intentions, best efforts, and hard work. These are the saddest entries. It *is* sad, indeed tragic, to see people slaving away, to the limits of their time and energy, but still failing to fulfil their aspirations. If hard work and best efforts were the answers, this would be a much fairer world. However, as we have seen in the previous chapter, theories are only useful if they apply to *this* world. The theory that best efforts will do the trick is, I fear, a theory for a different world. In this world, hard work at the wrong things not only fails to achieve: it makes things worse.

There is a silver lining to this cloud, however: it was Western Electric's discovery that hard work (at what turned out to be the wrong things) wasn't sufficient that led to Shewhart's fundamental work in the 1920s which in turn, as we know (Chapter 4), was the launchpad for Deming's own work that we are studying in this book. In Western Electric's case, as elsewhere:

> *"Best efforts*
> *are too often mere tampering,*
> *making things worse."*

We have studied this aspect of best efforts in Chapter 5; a common manifestation is the obsessive checking of all defects and failures, and attempts to react to everything which is found (see, for example, *Out of the Crisis*, pages 54–55).

Deming was asked to visit the Vice-President of a certain company. The Vice-President told him that, five years previously, he had been under the impression that best efforts, his and everybody else's, would produce quality and would ensure success. But not so. His company was eventually bought out by one of Deming's clients:

> *"Hard work and best efforts*
> *will by themselves*
> *not produce quality nor a market."*

So, lack of best efforts is not the problem. Deming often throws out a challenge to his audience:

> *"Who is not putting forth best efforts?*
> *Let him stand!"*

He continues to enquire, but as yet nobody has come forward to confess. Misdirected best efforts are a greater problem. We often hear these days of Deming's "Theorem Number 2"[2] which says that "we are being ruined by people doing their best." Best efforts are destroying us. It would be advantageous if people held back, if fewer people did their best:

> *"We would be better off if some people stayed in bed*
> *until it was too late to go to work!"*

Best efforts, not guided by theory, make things worse. A favourite saying of Deming's a few years ago, but less common nowadays, is that we need to "work smarter, not harder."

A personal thought on best efforts is that they are a further example of the hazards of suboptimisation. By definition, "best efforts" must optimise something, some tiny sub-system or process. Unfortunately, it is probably not important to optimise

---

[2] Theorem Number 1 is: "Nobody gives a hoot about profit." Think on!

that sub-system in its own right; of greater importance is that it contributes optimally to the larger system which contains it. And, as we have seen several times, particularly e.g. in the example at the end of Chapter 15, it is likely not only that these two objectives fail to coincide, but that they actually pull against each other.

Yet another of Deming's descriptions of the purpose of the four-day seminar is that:

> *"We are here to see*
> *how to make best efforts effective."*

Best efforts are not, of course, *inherently* wrong. They are wrong because the way in which they are used is wrong. They are wrong because something important is missing. And what is missing is a little ingredient which Deming has come to call "A System of Profound Knowledge" ...

Chapter 18

A SYSTEM OF PROFOUND KNOWLEDGE

---

*"What is needed? Profound Knowledge.
That's what these four days are for—
to learn something of it."*

Profound Knowledge. Could any other two words better describe the nature of the Deming philosophy? I doubt it. Yet Deming himself seems to have started using the term only quite recently: the expression is nowhere to be seen in *Out of the Crisis.*

When I first heard Dr. Deming refer to Profound Knowledge (I believe it was in 1987), I interpreted it specifically as the understanding of variation: common and special causes, tampering, the Rules of the Funnel, etc., including in particular their application not only to manufacturing processes but to processes of all kinds, especially as they affect people. By the summer of 1988, Deming was providing the following brief list of topics to be considered under the title of Profound Knowledge:

> Variation;
> Interaction of Forces;
> Operational Definitions;
> Psychology;
> Materials.

By the time of the British Deming Association's second National Conference in April 1989, this list had grown to eight items, now with concise and comprehensive descriptions instead of just the above rather terse headings. By early July, when Deming presented his four-day seminar in London, and then gave further presentations elsewhere in Britain and in France, the list had been further expanded, both in number and in detail. There were yet more additions in a paper (already mentioned in Chapter 15) delivered at a meeting of the Institute of Management Sciences in Osaka near the end of that same month.

Within just a few more weeks there was an especially significant development: Deming began to speak of a *System* of Profound Knowledge,[1] emphasising the interlinking and interdependent nature of the items and concepts presented in this chapter. By the end of the year, that System had been further shaped and honed.[2]

The first documented version of a System of Profound Knowledge was dated 3 January 1990. Substantially revised versions appeared on 3 March and on 11 May. After a pause, necessitated by a period of illness later in 1990, frequent further revisions were made during 1991. The present chapter has consequently been revised for the second printing of this book, in order to incorporate more of Dr. Deming's current thinking.

As with our treatment of the 14 Points (in Chapter 3 and Part 5), this chapter does not slavishly follow any particular version. It includes material based not only on the various editions mentioned above but also on other sources, particularly the presentation in Versailles which was mentioned in Chapter 17.

---

[1] I am grateful to Patrick Dolan, then the BDA's Secretary General, for first making me aware of this development when he returned from a brief visit to the USA in September 1989.

[2] I was greatly indebted to Dr. Deming for communicating these advances to me, as I had had no chance to meet up with him between the summer and the time when this book was completed.

A substantial proportion of the writing here appears in Dr. Deming's original words.

Naturally, many of the topics appearing in this chapter have already been discussed in some detail elsewhere in this book, and others will be addressed during Part 5. Consequently, our deliberations here will be relatively brief, so that the chapter will serve as a compact summary and reference,[3] much like Deming's allusion to a "bird's eye view" in the Introduction to Part 4, also providing a focus for our remaining more detailed consideration of the 14 Points with which we complete the book. However, there will also be glimpses of some deeper considerations to which the Deming philosophy inevitably leads.

The System of Profound Knowledge is described here in four Parts plus an Introduction and a concluding Summary, all interrelated. The four Parts are:

A. Appreciation for a system;
B. Some knowledge about variation;
C. Some theory of knowledge; and
D. Some knowledge of psychology.

INTRODUCTION

Knowledge that the prevailing style of management must change is necessary but not sufficient. We must understand what changes to make. There is in any journey an origin and a destination. The origin is the present style of management. The destination is transformation. The transformation will lead to adoption of what we have learned to call a *system*, and *optimisation* of performance relative to the *aim* of the system. The individual components of the system, instead of being competi-

---

[3] To aid this objective, there is considerable cross-referencing between this chapter and the rest of the book.

tive, will for optimisation reinforce each other for accomplishment of this aim. Such transformation is needed in industry, in government, and in education.

The journey requires Profound Knowledge as guide. As a good rule, Profound Knowledge comes from the outside, for a system cannot understand itself. Further, it comes by invitation, for Profound Knowledge cannot be forced onto anybody:

> **"A company that seeks the help of**
> **Profound Knowledge**
> **is already poised for the transformation."**

The journey to transformation requires *leaders*. How may a leader accomplish transformation?

First, he has a *theory*—a vision of his organisation as it would be when transformed. He understands why the transformation would bring gains to his organisation and to all the people with whom his organisation deals. Second, he is a *practical* man. He has a plan, and the plan is not so difficult that it cannot be carried out.

A leader must guide his organisation through the stages of transformation. But what is in his own head is not enough. He must understand people. He must possess persuasive power. He must convince and change enough people in power to make it happen. We call this person a *leader of transformation*.

No-one, leader or other, need be eminent in any part of Profound Knowledge in order to understand Profound Knowledge as a system, and to apply it.

> Indeed, specialism may be a hindrance. I (the author) realise now that my familiarity with "traditional" statistics closed my mind for a while to what Deming had to teach on statistical theory and practice. I know other statisticians are similarly obstructed. (I believe the same to be true of experts in systems and psychology regarding their related parts of Profound Knowledge.) If statisticians understood the difference between enumerative and

analytic statistical studies (see Item 7 in Part B), they would no longer teach and attempt to use techniques such as hypothesis tests and confidence intervals as methods of prediction.

As with any system, the various components of the System of Profound Knowledge cannot be separated: they are strongly interrelated (see Chapter 8). For example, knowledge of psychology is incomplete without knowledge of variation. A psychologist that possesses even a crude understanding of variation as learned in the Experiment on Red Beads (Chapter 6) could no longer participate in continual refinement of instruments for rating people.

A leader of transformation, and managers involved, need to learn the psychology of individuals, the psychology of society, and the psychology of change.[4] Anything less than this will fail to optimise over the long term the performance of the system.

Statistical theory, perhaps more than any other discipline, can contribute to improvement of management in industry, in government, and in education. Some understanding of variation, including appreciation of a stable system, and some understanding of special causes and common causes of variation, is essential for management of a system, including leadership of people. Theory of variation is helpful for understanding differences between people, interactions between people, and interactions between people and the systems that they work in and learn in. Statistical theory leads to better understanding of the evils of the prevailing system of management, and it indicates the way to better practice.

Statistical theory, used cautiously, with the help of the theory of knowledge, can be useful in interpretation of the results of tests and experiments. The interpretation of the results of tests and experiments is prediction.

---

[4] See, for example, the paper by S. L. Fink: "Crisis and Motivation: a Theoretical Model."

The theory of knowledge helps us to understand that management in any form is prediction.

> Grades given by teachers in school are judgments and rankings on *past* performance, but the grades are invalidly and cruelly used for prediction of *future* performance in another course or in a job. Likewise, appraisal of employees' *past* performance is used as prediction of their *future* performance.

The theory of knowledge teaches us that a statement, if it conveys knowledge, predicts future outcome (with risk of being wrong), and it fits without failure observations of the past.

Management of a system is action based on prediction. Rational prediction requires theory, along with systematic revision and extension of theory based on comparison of predictions with observed short-term and long-term results that arise from alternative courses of action. We learn this by study of the theory of knowledge.

A. APPRECIATION FOR A SYSTEM

In Chapter 8, a system was defined as "a network of functions or activities within an organisation that work together for the aim of the organisation." The importance of flow-charts in understanding systems was also stressed there; and, linked with this, was the comparison of the famous "Production Viewed as a System" diagram (Figure 23) with the traditional hierarchy, pyramid-style organisation chart (Figure 24).

Nida Backaitis has recently strengthened this material by pointing out that Figure 23 is the *real* organisation chart. For it shows people what their jobs are, how their work fits in with the work of others in the system, and thus how they should interact with one another as part of that system. The hierarchy diagram does none of this: it merely shows who reports to whom. In fact, if the hierarchy diagram conveys any message at all, it is the wrong message: namely, that a person's job is to please his boss (analogous to aiming for a high rating). It will probably be apparent (if not, there is more in Chapter 30) that viewing the hierarchy diagram as an organisation chart not only fails to aid optimisation of the system: it actually *destroys* the system (if ever one was intended).

A system of schools—public schools, private schools, parochial schools, trade schools, universities, for example—is not merely pupils, teachers, parents, school boards, and board of regents (using American terminology). It should be instead a system of education in which pupils from toddlers on up take joy in learning, free from the extrinsic motivators of grades and gold stars, and in which teachers take joy in their work, free from fear of ranking. It needs to be a system that recognises differences between pupils and differences between teachers (see Part D of this chapter).

What is the *aim* of a system? (For without an aim there is no system.) There are no theorems from which to derive the aim: the aim is a value-judgment.

The aim proposed here is for *everybody to gain over the long term* (it cannot all happen in the short term). "Everybody" includes employees, customers, suppliers, stockholders, the community at large, the environment. Examples of relevant and consistent aims are to provide good leadership for employees, opportunities for training, education for further growth, and other contributors to joy in work and quality of life. Again, anything less than this would fail to optimise over the long term the performance of the system.

The aim of the system must surely be consistent with the "constancy of purpose for continual improvement" in the first of the 14 Points. As we have seen in Chapters 8 and 15 respectively, the purpose of considering a system and the purpose of cooperation are one and the same: *they are to seek optimisation of the whole* rather than suboptimisation of parts. Optimisation is a

process of orchestrating the efforts of all components toward
achievement of the aim. Without an aim such as described
above, why *should* people cooperate to accomplish it? The aim,
whatever it be, must be made clear to everyone in the system.
Otherwise, how *could* people cooperate to accomplish it? Chapter 31 may aid the derivation of an appropriate aim for the system of education described above.

The components of a system could in principle, under stable conditions, manage themselves to accomplish their aim.

> A possible example is a string quartet. Each
> member supports the other three. None of them is there to
> attract individual attention. Four simultaneous solos do
> not make a string quartet. The four players practise singly
> and together to accomplish their aim. Their aim is
> challenge for self-satisfaction, and to provide pleasure to
> listeners.
> Recall that Profound Knowledge comes from the
> outside: any system needs guidance from the outside. The
> string quartet may well study under a master. Note though
> that the master need not be present at a performance.

The greater the interdependence between components, and
the larger the system, the greater be the need for communication
and cooperation between them (see Chapter 8). Also, the greater
is the need for overall management.

> A larger musical example is, of course, an orchestra
> (see Chapter 3). Thus, each of the 140 players in the Royal
> Philharmonic Orchestra of London is there to support the
> other 139 players. Listeners judge an orchestra not so
> much by illustrious players but by the way they work
> together. The conductor, as "manager", begets cooperation
> between the players, as a system, every player supporting
> the others. There are other aims for an orchestra, such as
> joy in work for the players and for the conductor.
> This illustrates the statement in Chapter 8 that the
> performance of any component needs to be evaluated in
> terms of its contribution to the aim of the system, not on
> any individual, competitive measure. A loss-leader in a

266

store is a further example. The store loses money on the loss-leader but supposedly makes great gains on the sale of other items to the hordes of people that are attracted by the loss-leader.

So, for optimisation, a system must be managed, the more so if there is growth in size and complexity of the system, and rapid changes with time. We have also remarked in Chapter 8 that an additional responsibility of management is to be ready to extend the boundary of the system to better serve the aim.

Thus it is wrong to suppose that changes in direction are a sign of wrong decisions and weak management. Competent management makes changes as required. The pilot, human or automatic, makes frequent changes in the course of an aeroplane. Without these changes, an aeroplane bound for Los Angeles might land in Burbank or Ontario. One could say that the aeroplane is never on course. Changes in wind, changes in temperature and in the density of the air (thin though it be), changes in traffic, all require changes of course.

This is not all. The destination may change *en route*, because of external forces that time may bring. It is management's job to provide change of course.

Management and leaders have still another job, namely to govern their own future, not to be merely a victim of circumstances. As an example, instead of taking loss from spurts in production to meet increased demand, followed by further losses from troughs of decreased demand, it might be better to level production, or to increase production at an economical rate. As another example, management may change the course of the company and indeed of the industry by anticipation of needs of customers for new product or new service.

The search for optimisation may be more rewarding than expected. To achieve the optimum may appear impossible; indeed, it may well *be* impossible, for a precise optimum would be difficult to define, let alone discover. But, recalling the Taguchi loss function (Chapter 11), the penalties for error within the vicinity of the optimum are tiny. With that loss function, one may move a short distance along the curve in either direction from the optimum but rise in the vertical only an imperceptible distance. So, fortunately, precise optimisation is not necessary: it will suffice to come close to it.

However, there is no doubt that suboptimisation is easier than optimisation. But it is also costly: it gives the impression of improvement yet, in reality, it builds barriers which obstruct real progress. Not only that: suboptimisation of one part often harms other parts so that, overall, the change causes more harm than good, besides making beneficial change more difficult.

Amongst the many examples of suboptimisation in the management of people are: the destructive effect of grading in school, from toddlers on up through the university; gold stars and prizes in school; the destructive effect of the so-called merit system; incentive pay; MBR; MBIR (see Dr. Deming's Foreword); and quotas. See also the Faulty Practices of Management in Chapter 29 and the story of the $138 saving in Chapter 27.

Yet more examples of suboptimisation through short-termism follow:

a. The quarterly dividend: make it look good. Aim for quick return on investment. High dividends, now: never mind the future.
b. Pension funds must be invested for highest apparent return. This requirement leads to rapid movement of huge sums of money—churning money out of one company into another.
c. Ship everything on hand at the end of the month (or quarter). Never mind its quality: mark it shipped, show it as accounts receivable.
d. Make this quarter look good. Defer till next quarter repairs, maintenance, and orders for material.

Failure of management to comprehend interdependence between components is in fact the cause of loss from use of MBO (as practised). The efforts of the various divisions in a company, each given a goal, are not additive: their effects are interdependent. One division, under pressure to achieve its goals, can kill off another division.

> Peter Drucker warned readers that overall results are not additive, that the results of various components of a company are almost always interdependent (see his book *Management, Tasks, Responsibilities, Practices*). Unfortunately, people that apply MBO do not heed his warning.

If optimisation needs cooperation, then competition must lead to suboptimisation. If economists understood the theory of a system, and the role of cooperation in optimisation, they would no longer teach and preach salvation through adversarial competition. They would, instead, lead us into optimisation, in which everybody would come out ahead.

Optimisation for everyone concerned should be the basis for negotiation between any two people, between divisions, between union and management, between competitors, between countries. Everybody would gain.

> The possibility of optimisation is voided if one party goes into negotiation with the avowed aim to defend his rights, or if he sets out with demands, and stands firm on them, with a time-limit for assent. The system of education described earlier would be destroyed if some group of schools band together for their own special interests, terms, or funds.

There are serious losses from competition for share of market and from barriers to trade. Often a company's main aim is just to get a bigger slice of the pie from another company. The same holds with nations. We have studied the harm caused by this approach in Chapter 15. If this is the sole or prime aim, the result is loss. The aim must be to bake a bigger pie, which brings more profit. Everybody wins on that. Trade barriers are barriers to progress. With them in place, there is no encouragement for producers of poor quality to improve since they then have a captive market, and there is no encouragement for producers of good quality to improve because their market is then restricted. That is Lose-Lose, not Win-Win:

"Where are your figures for these losses?"

B. SOME KNOWLEDGE ABOUT VARIATION

1. We need **knowledge about variation** because we live in a world which is full of variation.

> In this world there has always been variation, and there will always be variation—between people, in output, in performance, in service, in product. We must learn what the variation is trying to tell us. Should we try to do something about it and, if so, how?

2. Most losses are unknown, often unrecognised, not even suspected. We must learn to look out for the **two kinds of mistakes** (see Chapter 4), both of which cause huge losses, beyond calculation:

Mistake 1:

To react to any fault, complaint, mistake, breakdown, accident, shortage as if it came from a special cause when in fact there was nothing special at all, i.e. when it came from the system: from random variation due to common causes.

Mistake 2:

To attribute to common causes any fault, complaint, mistake, breakdown, accident, shortage when actually it came from a special cause:

"Where are the costs
from mixing up the two types of causes?
There are no figures.
How could the accountant know?"

Knowledge of variation helps us to understand these two kinds of mistakes and the losses they bring. Mistake 1 is tampering (see, in particular, Chapters 4 and 5); best efforts are often tampering, making things worse. The story related at the beginning of Chapter 4 is a supreme example of Mistake 2.

3. There is no way of always choosing correctly between the two types of causes, and there never will be. So we need **knowledge of procedures aimed at minimum economic loss from these mistakes.**

How can we aim for minimum economic loss? How can we judge a process to be stable or unstable? This has nothing to do with *probabilities* of the two kinds of mistakes. How could they be calculated? How could they even be defined when, as we have seen before, "no process ... is steady; unwavering" (*Out of the Crisis*, page 334)? Even if they did exist, how could they be used when most costs are unknown and, indeed, unknowable? What we need is an operational definition, a communicable rule of when to act and guidance on how to act. Shewhart provided it in 1924 (see Chapters 4 and 7). Many people still do not use it. Many that do still do not understand it.

4. We need understanding of the capability of a system and **knowledge about losses from demands that lie beyond the capability of the system,** demands often made through the mechanism of Management by Objective (see Chapter 29).

Figures on losses from MBO are also unknown and unknowable; but they are losses which cannot be tolerated. That which a (stable) system is capable of producing is described by its control limits. If a desired goal is outside these limits, the only sensible way to reach it is to change the system appropriately: management's job. Merely setting the goal is tampering. If we ask people to perform beyond the capability of the system, they can only meet the requirements by cheating. The goal often *can* be reached, but only by increasing variation and harming the performance of the system as a whole.

Note that it makes no sense to talk of the *capability* of a system when that system is unstable, i.e. unpredictable. Once a process has been brought into a state of statistical control then, and only then, does it have a definable capability.

5. It is necessary that statisticians **understand a system**, and how statistical theory can play a vital part in optimisation of a system. See also Chapter 8.

6. We need **knowledge about the interaction of forces,** including the effect of the system on the performance of people. Interaction of forces may work for good or ill: interaction may reinforce efforts, or it may nullify them.

    We need understanding of the dependence and interdependence between people, groups, divisions, companies, industries, countries. Unlike in past generations, anything that happens in the world now is of importance to us. We must be aware of dependence and interdependence, and the effect that it has on our work, our product, our service, our quality; otherwise we shall have no protection against the hazard of suboptimisation (see also Chapters 8 and 15).

7. We need knowledge about the **different kinds of uncertainty** in statistical data, and understanding of the distinction between enumerative and analytic problems.[5]

    We need knowledge about the different *sources* of uncertainty in statistical data. How were the data obtained? Were there errors in the sampling process? Were there built-in deficiencies? Were there blemishes and blunders in measurement, or in the interviewing process? What were the errors in responses, and what were the errors caused by non-response?

---

[5] See Chapter 7 of Deming's 1950 book: *Some Theory of Sampling* and his papers: "On the Distinction between Enumerative and Analytic Surveys" and "On Probability as a Basis for Action."

Statistical theory is vital for enumerative studies and in design of tests and experiments in medicine, pharmacology, the chemical industry, agriculture, forestry, and in any other industry. But the interpretation of results of a test or experiment is an analytic problem. It is prediction that a specific change in a process or procedure, or that no change at all, will be the wiser choice for the future.

8. We need **knowledge about loss functions, in particular the Taguchi loss function** (Chapters 11 and 12).

The Taguchi loss function helps us in two important ways. First, it teaches us that quality cannot be defined in terms of meeting requirements, conformance to specifications, or Zero Defects. In or out of specifications, there is always loss, and that loss may be decreased by reducing variation. Second, by considering the steepness of the Taguchi loss function, we are guided as to which quality-characteristics need the most immediate attention of management. They cannot work on everything all at once, so their job is to find and work on the most critical.

9. We need **knowledge about the production of chaos,** the wild results and losses that come from unfortunate successive application of random forces or random changes which individually may be unimportant.

We think in particular of lessons from the Funnel Experiment (Chapter 5). Suppose we have a process whose common cause variation, though always undesirable, is not too serious. Best intentions without knowledge can produce Rule 2, maybe Rule 3, and very likely Rule 4, with the effects we have seen in Chapter 5. If the common cause variation itself is serious, the effects will be quicker, or worse, or both.

Recall two examples from Chapter 5. First, there is the policy of worker training worker. The second, and worse, example is executives working together on policy, again with the best of efforts and intentions, working hard without the benefit of Profound Knowledge, only doing their best but making things worse.

The same holds true for government agencies and committees. Enlargement of a committee does not necessarily improve results. Enlargement of a committee is not a way to acquire Profound Knowledge.

Consequences of that observation are frightening. True, popular vote acts as ballast over a dictator, but does it provide the right answer? Does the House of Bishops serve the Church better than governance vested in the Archbishop? History leads to grave doubts.

C. SOME THEORY OF KNOWLEDGE

"We don't install knowledge—
I wish we could."

1. Almost every act in **management requires prediction**. Any *rational* plan, simple or complex, requires prediction concerning conditions, behaviour, comparisons of performance between procedures, materials, etc.

> A person needs a plan for how he is to get home this evening, and for that plan he needs prediction. For example, he predicts that his car will start and run satisfactorily, and he plans accordingly. Or he predicts he may have trouble with his car, and that may necessitate a different plan. As part of this, he may need prediction about buses or trains.
>
> As a further example, suppose we have been using Method A for a particular task, but we now have some evidence that Method B is better. Do we change to Method B? Not necessarily. If it is only a little better, the change may not be worth the hassle. There may be some evidence that Method B is a *lot* better but, if that evidence is not convincing, we may still decide to retain Method A, on the grounds that the change may cause more harm than good.

2. **A statement devoid of rational prediction is of no help to management.**

3. By definition, **there is no prediction in an unstable system;** in a stable system, prediction is provided by control limits. In either case, there is considerable difference between the past and future tenses when looking for what is best.

4. **Interpretation of data from a test or experiment is prediction:** what will happen on application of the conclusions or recommendations that are drawn from the test or experiment? This prediction will depend on knowledge of the subject-matter, not just on statistical theory.

5. **Theory leads to prediction.** Without prediction, experience and examples teach nothing (see Chapter 16).

6. **Theory leads to questions.** Without questions, one can only have an example. To copy an example of success, except when understanding it with the aid of theory, may lead to disaster. Something which may be excellent for somebody else, or in another place, may be destructive for us.

7. **No number of examples establishes a theory.** But a single unexplained failure of a theory requires modification or even abandonment of the theory (again see Chapter 16).

> The barnyard rooster Chanticleer[6] had a theory. He crowed every morning, putting forth all his energy, and flapped his wings. The sun came up. The connection was clear: his crowing caused the sun to come up. There was no question about his importance. There came a snag. He forgot one morning to crow. The sun came up anyhow. Crestfallen, he was forced to revise his theory. *Without his theory, he would have had nothing to revise, nothing to learn.*

---

[6] I am grateful to Brian Read for letting me know that Chanticle(e)r is the cock in *Reynard the Fox* (from Jean de la Fontaine's fables: *Fables Choisies Mises en Vers*, 1668).

Plain Euclidean geometry served the world well for a flat earth. Every corollary and every theorem in the book is correct in its own world. The theory for a flat earth fails when man extends his horizon to bigger buildings, and to roads that go beyond the village. Parallel lines on the earth are no longer equidistant. The angles of a triangle no longer add up to 180°. Spherical correction is required—a new geometry. Again, there is nothing wrong with the geometry for a flat earth—when we are on a flat earth. *It is extension of application that discloses inadequacy of theory, and need for revision or new theory.*

8. **Operational definitions put communicable meanings into concepts** (see Chapter 7). Communication and negotiation (as between customer and supplier, between management and union, between countries) require for optimisation operational definitions.

    We need to know precisely what procedure to use in order to measure or judge something, and we need an unambiguous decision-rule to tell us how to act on the result obtained.

9. There is **no true value** of any characteristic, state, or condition that is defined in terms of measurement or observation (again see Chapter 7).

    There is a number that we get by carrying out a procedure; change the procedure, and we are likely to get a different number. There are many examples in Chapter 7. As another: how would you count the boat people in San Diego?

10. There is **no such thing as a fact** concerning an empirical observation. Any two people may have different ideas on what is important to know about any event, and hence what to record concerning anything which has happened.

    "Get the facts!" Is there any meaning to this exhortation?

D. SOME KNOWLEDGE OF PSYCHOLOGY

1. **Psychology helps us to understand people**: the interactions between people and circumstances, interactions between teacher and pupil, interactions between a leader and his people, and the interactions in any system of management.

> Psychology helps us to understand (predict) how uncertainty and variation in circumstances affect people. Circumstances will affect people in different ways. The interaction of any person with circumstances may vary rapidly with time. The reward system in a company is an example of circumstances; the management is another.

2. **People are different from one another.** A leader of people must be aware of these differences, and use them for optimisation of everybody's abilities and inclinations. Management of industry, government, and education operate today under the supposition that all people are alike.

> Some care is needed in interpreting these statements. Who could deny that all people are different? But we must take care to differentiate between perceived differences in *effort* and *natural ability*. The second sentence above clarifies that Deming is concerned with the latter. It is a fact of life that there are huge natural differences between people. Should we just rue that fact because it makes management more difficult? It is surely more profitable to regard the differences positively, realising that potential for progress is greatly enhanced by recognising and combining different abilities and talents in a cooperative environment. "Variety is," after all, "the spice of life."
>
> Regarding the third sentence, probably most would deny that they manage under such a supposition (as regards natural ability). But much of what they do seems justifiable only under that supposition. Merit ranking, financial incentives, and bonuses are surely intended to encourage and reward *effort*. What could be the point of rating natural tendencies when the real gain is achievable by acknowledging and combining them?

3. **People learn in different ways,** and at different speeds (see Chapter 24).

> Some learn best by reading, some by listening, some by pictures (still or moving), yet others by watching someone do it. Some will need all four: that is no disgrace. How can there be any justification in penalising or punishing anybody for not doing what he has not been taught to do? People are not machines. They need to be treated with greater care, but they repay many times over the investment in that care.

4. **A manager, by virtue of his authority, has obligation to make change** in the system of management that will bring improvement.

> Since the major sources of difficulty and waste are the systems (common cause variation) within which people work, rather than the people themselves, the major responsibility for improvement therefore lies with those who have authority over systems, not with those who suffer from them.

5. **We must appreciate and value intrinsic motivation.**

> Intrinsic motivation is a person's innate dignity and self-esteem, and his natural esteem for other people. One is born with a natural inclination to learn and to be innovative (see the Life Diagram in Chapter 30). One inherits a right to enjoy his work. Psychology helps us to nurture and preserve these positive innate attributes of people.
> Modern management has stolen and smothered intrinsic motivation and dignity. It has removed joy in work and in learning. We must give back to people intrinsic motivation: for innovation, for improvement, for joy in work, for joy in learning. The need is to make a person responsible only to himself. Then he will learn: then he will produce new knowledge, innovation, new technology, new application of knowledge. All this has been smothered by modern management.

6. **We must beware the danger and damage of depending on extrinsic motivation.**

> Modern management is based on extrinsic motivation, motivation which comes instead from external reward.
>
> Extrinsic motivation is submission to external forces that neutralise intrinsic motivation. Under extrinsic motivation, learning and joy in learning in school are submerged in order to capture top grades. On the job, joy in work and innovation become secondary to a good rating. One knows not joy in learning. He knows not joy in work.
>
> Extrinsic motivation is a Zero Defects mentality. One tries to protect what he has. He tries to avoid punishment. It is humiliating and degrading: it is merely a day's pay for a day's work. But pay, above a certain level, is not a motivator. (Pay is surely not an *intrinsic* motivator, and it is often not as important an *extrinsic* one as most people think—see Frederick Herzberg's paper: "One More Time. How Do You Motivate Employees?") We must get back to the individual, and give him joy in what he needs to do.

7. **We must also avoid overjustification.**

> Overjustification comes from faulty systems of reward. Overjustification is resignation to outside forces. It could be monetary reward to somebody, or a prize, for an act or achievement that he did for sheer pleasure and self-satisfaction. The result of reward under these conditions is to throttle repetition: he will lose interest in such pursuits: he may never do it again.
>
> In this regard, Dr. Deming often cites the mother who paid her little boy for voluntarily washing the dishes: "He never washed another dish"; and he himself for having tried to tip people who help him in airports or hotels. The offered money was regarded as an insult: "He had carried the bag for *me*, not for *pay*." But *refusing* to tip is often regarded as an insult. Why the paradox? In the former cases, the strength of human relationship makes the money worse than an irrelevance; in the latter, the money is the only relationship. Strengthening human relationships and respect is surely part of the transformation.
>
> Monetary reward is a way out for managers that do not understand how to manage intrinsic motivation.

SUMMARY

There is **necessity for transformation in government, in industry, and in education** from Competition: Win-Lose to Cooperation: Win-Win with no loser. Again, this is transformation back to intrinsic motivation, to the individual, back to where we were 50 or 100 years ago. Let us give the individual a chance again. It is also transformation back to leadership of people (Chapter 25). It is time that we all cooperated more (Chapter 15) and learned more from each other. We need to change with better understanding, with the System of Profound Knowledge which will show us the road and the roadmap. There is no other way. But it is purely a matter of choice, not force. No government can order it. It has to be voluntary.

This is Profound Knowledge—some of it. It is not just management that have a new job; educators have a new job:

> *"This should be the backbone of courses*
> *in Schools of Business*
> *and in Departments of Statistics.*
> *This is what they ought to teach*
> *—and will do two decades from now—*
> *if we're still around then."*

Deming also claims that much of this System of Profound Knowledge has "been known for generations." But in the welter of change and pressure, much of it self-inflicted, we have ignored it and forgotten it. As Deming learned in 1923 from Dr. Alderson, Professor of English and Dean of the School of Mines at the University of Colorado:

"THERE IS NO SUBSTITUTE FOR KNOWLEDGE."

We continue to ignore and forget that at our peril.

PART 5

THE 14 POINTS REVISITED

THE 14 POINTS REVISITED

And so we return almost to where we began: Dr. Deming's 14 Points for Management.

Having reached this stage of the book, the reader should be on his guard against false impressions from which some people suffer. Let's be clear. We can never "do" the 14 Points, not in any complete sense. Even when some of the words seem to have been obeyed, there is always more to do in the areas on which the Points focus our attention. In any case, we know that there is no fixed, standard version of the 14 Points. When Deming feels that he can improve on any of them, he does so. And he *has* done so several times. A number of the amendments which have been made were identified in Chapter 3. There have recently been further alterations, though they have been more for establishing a clearer focus than in the nature of major change.

The 14 Points are a substantial aid to teaching and to learning. But, although the 14 Points are an extremely helpful guide to important parts of the philosophy, they may also increase the danger of people trying to take action in order to obey words before developing understanding. That can be a fatal mistake, and I warned against it in Chapter 3. The context of the 14 Points is a commitment to continual improvement in quality, in its widest sense and interpretation, and what is needed to fulfil that commitment. That certainly involves a lot of action—but it also involves a lot of education and understanding of why that action is needed, and of the pathways that need to be cleared before some of the action becomes appropriate.

I know a number of sad tales about how, naturally with the best of intentions, over-enthusiasm to plunge into "implementing" the 14 Points has had very damaging effects. If you view the *Doctor's Orders* video, you will hear me prompting Deming, in a somewhat puzzled way, to reiterate a story that I had recently heard him tell at a four-day seminar in London. I sounded puzzled because I was uncertain that I had heard the story correctly. It seemed incredible. But Deming confirmed that there was indeed a manager in Ford who, after attending a four-day seminar, went back to work the following Monday and fired all his inspectors! He thought that that was what Point 3 was telling him to do. Or perhaps the truth is that he didn't really think at all. What better reminder could we have that it is impossible to *install* knowledge?

Hasty action, such as that demonstrated by the Ford manager, and the general tendency to jump to conclusions based on insufficient understanding, can harm the reputation and credibility of the Deming philosophy itself. Consultants and teachers and other workers in the field must be aware of the dangers. If you listen carefully to Deming, you will soon find that he has a (to some) unexpectedly practical and pragmatic approach to issues on which he is often criticised. He is sometimes misquoted and often misinterpreted as saying things that are totally unrealistic. But we have already pointed out how so many of his "unrealistic" utterances in the early 1980s have now become fact, or are well on their way to doing so—and there was also, of course, that apparently unrealistic claim in 1950 that Japan would become a major industrial power. Beware of dismissing Deming's thoughts because some of what he says appears impossible in today's world. Thought only of what is possible or impossible *now* immediately falls prey to the second Deadly Disease, the over-concentration on short-term thinking.

Of course, we have to deal with the short term according to the circumstances in which we find ourselves. But are we resigned to always staggering from one short-term crisis to another without having any semblance of a long-term direction to look toward? Naturally, we can't do everything now which favours the long-term direction rather than the short-term facts. Deming doesn't ask us to. He does not call for sudden abolition of mass inspection schemes before things have been improved sufficiently to render them unnecessary. He does not advocate immediately cancelling piles of contracts so that we can abide by the single-supplier principle. Indeed, he *agrees* that there will be many cases in which the single-supplier situation may never be reached. And he well realises that even performance appraisal, which everyone knows he despises, cannot be scrapped overnight. As he says on page 88 of his 1982 book:

> **"A big ship, travelling at full speed,**
> **requires distance and time to turn around."**

But he does, quite rightly, expect us to make a start straightaway on learning and understanding *why* the 14 Points are what they are. He does expect us to begin as soon as possible to lay the groundwork which will enable the right changes to be made. He does expect us to commence making constructive moves in the right direction as soon as that becomes feasible.

While we're about it, let's lay a few other ghosts to rest. Deming does not say in Point 4 that the price of supplies doesn't matter. And he does not say in Point 11 that management does not need figures to manage with, and that neither they nor their workers nor their company need any goals or objectives. And (Point 10) he is not against the use of *all* posters and slogans. Nor (in Points 1 and 5) does he expect us to continually improve everything all at once.

But let's not swing the pendulum too far the other way. Agreeing that we can't do everything in the 14 Points at once is not letting us off the hook—it is simply being practical. There is no implication here, explicit or implicit, that anything in the 14

Points is unimportant and can be ignored. In fact, I recall one four-day seminar at which Deming hit the audience straight between the eyes at the very start of discussion with:

> ### *"Don't say you're going to obey some of them but not others."*

There is a world of difference between, on the one hand, holding fire on some of the more difficult aspects of the 14 Points because the right foundations have not yet been laid and and, on the other, planning to ignore them because you disagree with them. For the Points are not just ends in themselves. Each one is a vital component within the broad framework of the Deming philosophy. And these components do not stand alone: they are inextricably interlinked and overlapping with each other and with countless other aspects of Deming's teachings. Let us recall that

> ### *"The 14 Points for Management in industry, education, and government follow naturally as application of the System of Profound Knowledge."*

You just cannot throw away one of the 14 Points and kid yourself that you are left with thirteen-fourteenths of the Deming philosophy. In any case, that would imply that all 14 Points constitute *fourteen*–fourteenths, i.e. 100%, of the Deming philosophy— and we know well that that isn't so. But they *are* a good start! And absence of any Point incalculably weakens the whole structure.

Chapter 19

POINT 1: CONSTANCY OF PURPOSE

> **Create constancy of purpose for continual improvement of products and service,** allocating resources to provide for long-range needs rather than only short-term profitability, with a plan to become competitive, to stay in business, and to provide jobs.

An interesting variation which Deming used on one occasion referred to "constancy of purpose to help man live better materially." It crossed my mind that he could have added " ... and mentally."

How can such constancy of purpose be achieved? Surely, only by developing understanding of the need, and by understanding how continual improvement fulfils and satisfies that need. That is good and necessary theory. But how does Deming suggest it can be attained in practice? In schools, colleges, and universities: by joy in study; in industry: by joy in work. And a company neither deserves, nor stands any reasonable chance of getting, management and employees who have joy in their work unless that company has good, clear, and proper long-term aims and principles, genuinely held, with their employees being made fully aware of them and believing in them:

> *"Create and publish to all employees a statement of*
> *the aims and purposes of the company."*

> *"The values and beliefs of the organisation*
> *as set forth by the top management*
> *are important."*

(See also Deming's definition of a "system" in Chapter 8.)

The importance of Point 1 is stressed by the fact that it is the only Point to have a direct counterpart in the Deadly Diseases. Deadly Disease Number 1 is *lack* of constancy of purpose (see Chapter 3). Short-termism is, by definition, lack of constancy of purpose. With lack of constancy of purpose, nobody is really sure why the company is in business, or what its aims are, or what the future is likely to hold in store. This creates instability— and instability, as we know, increases variation and thus reduces quality:

> *"It is impossible to work for a company*
> *without knowing the aim of the company—*
> *and it must be constant,*
> *not changing with a new manager, a new President.*
> *Nobody can work otherwise."*

Of course, every company has to pay attention to the short term; we know that. Hopefully, most pay some attention to the medium and long term as well. But how much?—

> *"We hope for quick results,*
> *but we must not expect them."*

Where in the spectrum is the company's real focus? Is it primarily on the short term, so that the long term gets ruled by the way the wind happens to blow? If so, then management should be honest about it in order that everyone knows where he stands. Then there will be less effort and enthusiasm wasted on long-term planning which stands little chance of being used. Planning requires prediction, and prediction requires stability. Or is the focus on the long term—sure, subject to necessary adjustments to overcome short-term hazards, but with sights firmly set on the direction leading to continual improvement, with the aim to survive, to succeed, to serve customers well—and increasingly so—and, more broadly, to be of service to the country and indeed to the world? Does that sound "over the top"? Well, there are good precedents. Take, for example, the Taguchi loss function, discussed in Chapters 11 and 12. The "loss" in Taguchi's loss function is often expressed in terms of no less than the "loss to society" caused by not hitting nominal, rather than just of localised financial loss.

Bill Latzko[1] tells me that, in a version of the 14 Points provided to him by Dr. Deming, this broader focus was demonstrated by the addition of those same two words: " ... continual improvement of ... service *to society*." A vital role of business is, as the wording of the Point concludes, "to provide jobs." This is denied, regrettably, by smaller-minded, cost-cutting views of business demonstrated by some companies and indeed some governments. But if business does not provide jobs, who will? Let us absorb the fact (from the Preface of *Out of the Crisis*) that:

> "It is no longer socially acceptable to dump employees on to the heap of unemployed. Loss of market, and resulting unemployment, are not fore-

---

[1] Bill Latzko is a consultant who is especially well-known for his work in non-manufacturing areas, and also for his book: *Quality and Productivity for Bankers and Financial Managers*.

ordained. They are not inevitable. They are
man-made."

This is neither a suggestion nor an excuse for over-manning. It is a
further call on management not merely to increase the company's
market share but to innovate in order to expand the market.
Management need to develop the objective, and have the appropriate foresight, to design, produce, and deliver products and service which will prove enticing to customers and will thus build
markets and hence provide more jobs:

> *"It is a matter of optimisation—*
> *management for greatest service, maximum profit,*
> *and the best deal for everybody:*
> *employees and customers."*

Again, Deming's position is not on the left but, if anywhere, in
the center of what may be referred to as the "political divide,"
seeking good from both sides. Ne'er the twain shall meet? They
need to.

Now, let's not get carried away. We must not be like that
poor Ford manager who fired all his inspectors (see the Introduction to this Part of the book)! Yes, we do need to plan for the
long term; yes, we do need to think more of market expansion; yes,
we do need to be able to provide more jobs. Some people learning
of the 14 Points for the first time protest that they have to carry
on living in the world as it is, here and now. Well, of course: we'd
be pretty silly to spend all our time planning for a better tomorrow while forgetting to ensure that we survive today—with the
result that we would not be around to bring that better future into
existence!

That's not news to Deming. Right at the beginning of his
discussion of Point 1 (*Out of the Crisis*, page 24) he talks of the
two kinds of problems: "problems of today" and "problems of

tomorrow." But he also points out the danger of staying "bound up in the tangled knot of the problems of today." An important task of management is to exercise good judgment on allocation of resources and effort to the two sides of this picture. If everything, or almost everything, is devoted to the problems of today—patchwork, stamping out fires, cosmetics—then decline is guaranteed. The competition will not stand still; it knows that improvement means more than cosmetics. We must of course continue to deal with the emergencies, solve the problems, and stamp out the fires—but recall from Chapter 4 that putting out a fire in a building, a fire which has been ignited by the problems of today, does not result in any *improvement* to that building—it just saves it from being damaged or even totally destroyed right now. It is worth a thought though that the fire, and today's problem that set it, might not have been there for us to fight had we concentrated more on improvement yesterday.

Deming says that he first published such thoughts back in 1936, and that they are mere echoes of problems stated by George Edwards, one of those whom he numbers amongst the great men from the Bell Laboratories with whom he worked "as apprentice" in those far-off days.

Short-termism, lack of constancy of purpose, implies lack of job security. A management whose values are so shallow as to take on people only when they are forced to, and dispose of them just as thoughtlessly as if they were some inanimate commodity, should expect shallow treatment in return from those unfortunate enough to be employed in their company. Incidentally, I have never liked the term "human resources," especially since Del Nelson of the McClellen Air Force Base pointed out to me that what we do with a resource is to:

> obtain it;
> shape it;
> use it; and
> throw it away.

Finally, having stressed the importance of the provision of jobs, it's worth saying a few words on job descriptions. Deming considers that most job descriptions are woefully inadequate:

> *"A job description*
> *should not only describe what the job is,*
> *but also what it is for."*

And why? Because:

> *"Anybody's job*
> *should not merely be*
> *to 'do it right,'*
> *but to do it better."*

This is not new. Deming refers back to a paper, again written some 50 years ago, in which he said:

> *"Nobody is doing his job right*
> *unless he is continually collecting data*
> *to help improve that job."*

An interesting variation on this theme came from a presentation by Nissan at the Second Annual Conference of the British Deming Association, held in Portsmouth, England in April 1989. Nissan apparently use very little detail in their job descriptions: an engineer, for example, should be able to "do everything that an engineer needs to do"! Maybe it would be useful to start thinking in these terms as regards managers, statisticians, teachers, psychologists, consultants, and all that have contributions to make to the transformation.

Chapter 20

POINT 2: THE NEW PHILOSOPHY

Adopt the new philosophy. We are in a new economic age, created in Japan. We can no longer live with commonly-accepted levels of delays, mistakes, defective materials, and defective workmanship. Transformation of Western management style is necessary to halt the continued decline of industry.

There is little to write specifically under this Point. For this whole book is about the new philosophy and the needed transformation. Were there anything fundamentally new to add here, the rest of the book would be flawed.

What does the new philosophy imply? Some of the most important issues raised elsewhere in the book are that we are concerned with a new kind of economics, with a new system of reward based on cooperation as a system (rather than as an accident or as isolated efforts) instead of competition which can be ruinous; we are concerned with new thinking on leadership of people; and we are concerned with developing joy in work, and joy in cooperation with others who take joy in their work, as prerequisite to achieving what needs to be achieved.

For survival, let alone success, there is need for change, and that change is to adopt the new philosophy. What is the new philosophy? It is the philosophy of continually-improving quality and productivity long-term and forever. Note that this is therefore not separate from Point 1 but indeed very much overlaps with it. The backbone of this philosophy is transformation to the culture of Cooperation: Win-Win.

What does adopting the new philosophy mean in practice? It means the breaking down of many strong, thick barriers to improvement. Breaking down those barriers implies transformation of culture. The barriers include:

1. Unwillingness to change;

2. Fear of failure;

3. Fear of the unknown—"Where would change leave me?";

4. People measuring productivity rather than helping to improve it (Deming says there are far more of the former than the latter!);

5. Financial people who merely beat down costs rather than learning the new philosophy and helping accomplish the changes that must take place; and

6. The system of reward (see Chapter 30):

> ## *"The new philosophy*
> ## *is a new system of reward."*

In its context, the transformation, the metamorphosis, the change of state has been likened to the transition in high-jumping from the Western Roll to the Fosbury Flop, a technique due to Richard Fosbury who became Olympic champion in the Mexico Games of 1968. Before Fosbury, was there anything wrong with the Western Roll? No, it was the best-known approach. Would it be wrong now? Yes: it is obsolete, it cannot compare. I leave the reader to draw the analogy with the situation in management.

Incidentally, do you think the experts in the Western Roll were pleased when the Fosbury method arrived? It is hard when the approach with which you are familiar, in which you may be an expert, becomes redundant, out-of-date, pointless. What should management do? Keep on with the Western Roll, or start to learn the Fosbury Flop?

Chapter 21

POINT 3: CEASE DEPENDENCE ON MASS INSPECTION

Eliminate the need for mass inspection as the way of life to achieve quality by building quality into the product in the first place. Require statistical evidence of built-in quality in both manufacturing and purchasing functions.

Is Deming saying that we should abolish inspection? No, of course not. But are we still *depending* on mass inspection as the way of life to try to ensure that the customer gets some kind of quality? If so, then substantial change is sorely needed. For that is outmoded, except in sensitive life-and-death matters.

An article in *The Christian Science Monitor* in 1981 quoted Deming describing American management as "just burning toast and scraping it." Deming used to enjoy showing an advertisement which proclaimed:

> "Only M—— employs dozens of testers just to test its nozzles."

Maybe only M—— needed to! He interpreted the advertisement as a plain admission that they could not make them right. Apparently about one-third of that company's employees were involved in such testing. Just think of the additional cost which was incurred. It is an unfortunate sign of the times that this advertisement probably did increase sales. We are so used to suffering bad quality that 100% inspection sounds good.

I know of a similar story involving companies that manufacture floppy disks. One company improved its quality so much that it had no need to continue with mass inspection. A competing company, making inferior disks (so I am told), advertised 100% inspection. What do you think happened? Yes—the same sad sign of the times: sales of the inferior product went up, and those of the better product declined. Customers also need to learn to optimise rather than suboptimise; and, in time, they will.

Of course, we do not look to eradicate *all* inspection: it is *mass* inspection which should go. Mass inspection is costly, wasteful, nonproductive; it aims to sort out good from bad; it does not contribute to progress. There is the world of difference between, on the one hand, dependence on inspection as an attempt to provide the customer with something that he won't complain about and, on the other, the use of inspection to provide guidance toward improvement of a stable process as well as to pick up the occasional special cause that creeps unannounced into that otherwise stable system.

Dependence on mass inspection is fraught with danger and with high cost. In this day and age, it is a formula for going out of business. 100% inspection is very expensive, and we shall soon see some evidence that 100% inspection is not necessarily 100% effective. Less than 100% inspection immediately introduces the outmoded concept of AQL (Acceptable Quality Level). The AQL represents a supposedly-acceptable level of defectives —an immediate contradiction to the philosophy of continual improvement. The AQL concept is embodied in the use of the well-known American Military Standard 105D and of the similar British Standard BS6001.

The better way is, of course, to build quality into the process and product in the first place, the results being higher quality, lower price, bigger market share, bigger market. Actually, the two statisticians whose tables have often been used for such "acceptance sampling" procedures, H. F. Dodge and H. G. Romig, were *not* altogether out of tune with Shewhart's and Deming's thinking; to quote them:

> "Quality control is achieved most efficiently, of course, not by the inspection operation itself, but by getting at causes."

Would that more users of their tables heeded that observation!

If a Quality Control Department's job is merely that of ingoing and outgoing inspection (see *Out of the Crisis*, page 31), then it belongs to a past age. But QC Departments *are* often regarded in this way, both by themselves and others, and this is one reason why Deming speaks so disapprovingly of them. This is another reason why quality must come to be seen as everybody's job, not that of just one department.

Of course, Point 3 is one on which much progress has been made during recent years. A decade ago, many people could not conceive of how quality standards might be met without mass inspection. We suggested in Chapter 3 that some might laugh at the very thought. That laughter is still audible, but it is now considerably muted. However, I am obliged to my good friend John S. Dowd[1] (a consultant in California) for pointing out to me that mass inspection still exists widely—yet is unrecognised as such—in nonmanufacturing processes. For examples, are not

---

[1] I first met John Dowd while working with the British subsidiaries of the Nashua Corporation in the early 1980s. I have learned much from him, both then and subsequently.

auditing, validating reports, performance appraisals, and proof-reading all cases of mass inspection? During a visit to London in 1989, Deming showed an example in which processes, which had previously been beautifully stable, went berserk on the occasion of the auditor's visit.

It is not commonly realised how fallible inspection procedures themselves can be. The apparatus used for inspection often gives more trouble than the apparatus used for production. And very many inspection processes, upon investigation, show phenomena such as "flinching," illustrated in Figure 24 on page 266 of *Out of the Crisis*. This results from the attitude of "in case of doubt, pass it." Such an attitude is easily understandable in management environments ruled by quotas (c.f. Point 11) rather than by consideration of the customer.

The following striking illustration of human fallibility in inspection processes has been in circulation for at least 50 years. If you haven't seen it before, try it out for yourself right now. Suppose that the letter F represents a defective item, and all other letters good items. Don't look right now, but if you turn the page you will find displayed there a straightforward 17-word sentence; it begins with the word: "FINISHED ... "—there you are, one defective item straightaway! Give yourself, say, 15 to 20 seconds to read through this sentence and count the number of Fs. Easy enough? OK, check your watch, turn over the page, and try it!

How many Fs did you find? For the answer, and brief discussion, see page 305.

This was not just an academic exercise. Jerry Langley, of the API (Associates in Process Improvement) group, tells of a jacket he bought which contained in one of the pockets a number of inspection tickets. There were eight of them altogether. The list went something like this:

> Impression—inspected by Number 8
> Armhole pressing—inspected by Number 10
> Lining press—inspected by Number 16
> Lapel—inspected by Number 4
> Labelling—inspected by Number 6
> Blocking—inspected by Number 4
> Examiner's Inspection—Number 52
> (unspecified)—inspected by Number 9

Also, on those last two tickets:

> If any defect should be discovered,
> return this ticket with garment.

(I wonder what should be done with all the other tickets.) But what could possibly go wrong when the customers are being protected by that huge inspection process? Well, I know Jerry soon found that the lining was loose and that one of the pockets was not sewn at the bottom. Actually, it was the discovery of all those tickets that persuaded him to take a careful look for flaws! It is foolish to conclude, just because something is (said to be) inspected, that it must be all right. Of course, the label or sticker "QC passed" gives the *impression* that all must be well. You can believe it if you wish.

So it's not just an artificial problem of counting Fs: real inspection processes can also be very fallible. And just look what it costs to involve all those people in such inspection processes rather than in productive work.

This illustration falls foul of another problem which is highlighted by Deming: the fallacy of *divided responsibility*. Agreed, each inspector supposedly had a specific type of fault on which to focus (except that it appears there were also a couple of general checks). But what a soul-destroying task! And, if you know that seven other people are also going to be inspecting that garment, are you really going to devote much care and attention to your particular task?

FINISHED FILES ARE THE RESULT OF YEARS
OF SCIENTIFIC STUDY COMBINED WITH
THE EXPERIENCE OF MANY YEARS.

---

* * *

Two examples of the problem of divided responsibility in
Out of the Crisis are the late David Chambers' story of 11-fold
proof-reading (page 30) and the two signatures required on payroll cards (pages 208–209). Two further cases are often mentioned
in seminars. One is an operation which involved a "picker" and a
"checker"—one person picked items off the shelves in order to fill
requisitions: a second person then checked that the right items,
and in the right quantities, had been selected. Management
wondered why so many errors were being reported by the customers of this operation. Deming told them that the system guaranteed errors. The picker should become responsible for the
whole job: the job could be done properly by one person, and should
be. The management followed his advice; and the error-rate
plummeted. As Deming remarked: "It's all very simple."

In the other example, of particular interest because of a
widely reported near-miss between two aircraft, it turned out
that the FAA (Federal Aviation Administration) had ruled that
the pilot and co-pilot should both read the coordinates. Again,
either one could do it and get it right. It's straightforward; it's
not a matter of opinion. But *divided* responsibility is *reduced*
responsibility, a guarantee of error. Deming's thinking here is
that, when there is such duplication of effort, nobody has ownership of the job—so there is a natural tendency to be slipshod
through lack of *individual* responsibility and ownership for the
task.

It may have crossed the reader's mind that the Experiment on Red Beads (Chapter 6) seems to provide a counter-example to Deming's teaching on divided responsibility. Counting the number of red beads is a task apparently shared by (i.e. with responsibility divided between) two junior inspectors—yet Deming often says the inspection process is the only thing carried out *correctly* in the experiment. But there is a crucial difference between this situation and the ones so far described. In the Experiment on Red Beads, the counting is not actually *shared* by the two junior inspectors: they carry it out quite independently of each other. They do not see, check, or comment upon each other's findings; a third party, the Chief Inspector, is the only one to make comparisons. It is this mutual independence which is important, and which makes this inspection process a good one. This is a *valid* use of 200% inspection. 200% inspection can, contrary to all the other cases above, give net gain if administered properly. A second opinion, obtained *independently* from the first, can be valuable. The hazards of divided responsibility would be illustrated if instead, say, one of the inspectors counted the red beads, and then told the other the result he had obtained, asking him if he agreed. Consensus is a poor indicator of "rightness" (c.f. Rule 4 of The Funnel in Chapter 5). There is more to read on the problems arising from consensus on pages 442–444 in *Out of the Crisis*.

Until things improve very dramatically, some products will continue to need 100% inspection—and very effective 100% inspection at that. Matters of life and death, such as tubes for nuclear reactors, are a case in point. Chapter 15 of *Out of the Crisis* shows that, if processes are in statistical control, there are really only two choices: no inspection (except for other purposes as described above) or 100% inspection. *If processes are in control, a sample from a batch contains no information concerning the uninspected items in that batch.* It is to be noted that the choice between the two alternatives—whether to inspect or not—is made on the basis of economics: profits and losses; probability calculations such as those involved in acceptance sampling schemes are not relevant. Of course, if processes are out of control, many of the calculations in Deming's Chapter 15 cannot be carried out, as there is then no meaningful figure for the average fraction defective. In that case, referred to as a "state of chaos," 100% inspection deserves consideration. However, Joyce Orsini's rules[2] (*Out of the Crisis*, page 415) have been found to almost always save cost compared with 100% inspection.

As some final thoughts on inspection, Deming carries with him a four-point card which he picked up from an unnamed company. Here are the details (*with his comments*):

> Quality is expected by our customer.
> (*"Well, no fooling."*)

> Quality is the prime responsibility of the operator.
> (*"It is **not**!"*)

> The inspector shares this responsibility.
> (*"Can you imagine anything worse?"*)

> The inspector will use MIL-STD-105D to determine the acceptable number of defective items.
> (*No comment.*)

Incidentally, it has recently been announced that the American Department of Defense is discontinuing its use of MIL-STD-105D.

---

[2] Joyce Orsini's rules were developed in her Ph.D. thesis, written under Dr. Deming's supervision at New York University.

The answer is 6 Fs. If you don't believe it, turn back and check again.

I often use this example in seminars, and I reckon that there is only about a 25% success-rate. So, if you got it wrong, feel no shame—be sure you're in the majority! Answers usually range from 3 to 7, with the occasional 2 or 8. A full discussion of this experiment is given on page 27 of Bill Scherkenbach's book, *The Deming Route to Quality and Productivity*.

Chapter 22

POINT 4: END LOWEST–TENDER CONTRACTS

End the practice of awarding business solely on the basis of price tag. Instead, require meaningful measures of quality along with price. Reduce the number of suppliers for the same item by eliminating those that do not qualify with statistical and other evidence of quality. The aim is to minimise *total* cost, not merely initial cost, by minimising variation. This may be achievable by moving toward a single supplier for any one item, on a long-term relationship of loyalty and trust. Purchasing managers have a new job, and must learn it.

Point 4 is one concerning which some hasty people jump to wrong conclusions. They may proclaim that Deming says we don't need to bother about price when purchasing. Others just scoff at the whole single-supplier concept. These days, Deming often tackles the potential criticisms before they are even raised:

> *"Who would buy a tire for his automobile*
> *at the lowest price?"*

and:

> *"If you had, say, six suppliers,*
> *all pretty much the same as each other,*
> *would you buy on the lowest price?*
> *Yes, of course you would; you'd be a fool not to.*
> *But—that's not this world."*

Recall our earlier discussion on *theory* (Chapter 16). When considering any decision, we should be clear about the theory under which that decision makes sense. Any theory is correct in some localised context. But how localised is that context? Is it anywhere near the real world in which we live and do business?

To aid our thoughts on such matters as single-sourcing and buying on lowest tender, Deming has begun to speak of various "worlds"—a "world" being, in effect, what we have just referred to as a "context." World Number 1 is defined by the following conditions:

1. The customer knows his needs (as opposed to just thinking that he does) and can make clear what they are in operational terms.

2. There are, say, six suppliers that can meet his needs *exactly*—there is no variation. There are no differences between the suppliers except the prices that they are asking.

3. The purchase price of the supplies is the only relevant cost. Because of Condition 2, the cost of use is the same in each case.

In World Number 1, the customer should surely buy on lowest price. What is there for him to lose? World Number 1 is a theory. How useful is it? Answer truthfully!

So let's try World Number 2[1] in which:

1. The customer knows his needs and can make them clear in operational terms.

2. The six potential suppliers can all satisfy the customer's needs, which are expressed in terms of meeting specifications.

3. The costs of use will differ (since Condition 2 implies that there *is* variation in World Number 2).

To maximise profit in World Number 2, we need to consider not just purchase price but *total* cost: purchase price plus cost of use. Of course, that can be difficult, especially remembering all those unknowable figures which Lloyd Nelson warns us about and which, warming to the theme, Deming points out are unrecognised, not even suspected. It will require knowledge, study, and maybe experimentation.

World Number 2 is also a theory. How useful is it? How close is it to our world? Well, it's surely closer than World Number 1. What do we do? Just because World Number 2 is more difficult to deal with, do we pretend we are in World Number 1 and behave accordingly? What is that but another "lazy way out"? Is that the way to manage?

A particular complication in World Number 2, compared with World Number 1, is the expression of customer needs in terms of specifications. For we have seen in Chapters 11 and 12 that cost of use can vary enormously even if everything is within specifications. But, in some circumstances, that may not be a problem; life doesn't always appear to be so difficult. So consi-

---

[1] Dr. Deming's Worlds 2 and 3 are interchanged here for convenience.

der World Number 3, in which:

1. The customer needs a commodity, such as one-spoonful packets of sugar, where there is no problem with specifications. This product is only made by two national manufacturers, both eminently trustworthy.

2. The customer cannot buy directly from the manufacturers; however, there are six middlemen, all eager for his business. From Condition 1, we see that they all furnish the same commodity.

3. As Condition 2 seems to imply, the purchase price is all that matters.

Is it? What about service? If either one of them says that he will deliver it on Thursday at 3:00 pm, does he mean *this* Thursday? What about cleanliness of their vehicles, affecting cleanliness of the delivered product? What about accuracy and timeliness of their paperwork? What about their help with the unloading? No, Conditions 1 and 2 of World Number 3 may be close to your real world, but Condition 3 is not.

     Distrust of the policy of awarding business on lowest price is not new. Quoting from John Ruskin, the nineteenth century English art critic:

> "It's unwise to pay too much, but it's worse to pay too little. When you pay too much, you lose a little money—that's all. When you pay too little, you sometimes lose everything because the thing you bought was incapable of doing the thing it was bought to do."

and:

> "There is hardly anything in the world that some man cannot make a little shoddier and sell a little cheaper, and people who consider price only are this man's lawful prey."

See also the statement reported by Jeanne Perreault (*Out of the Crisis*, page 33).

Let us now turn to Point 4's other bone of contention: single-sourcing. I remember Deming being asked if we should always opt for a single supplier on any one item. He replied:

"No—you have to be practical."

Note that the wording of this Point has been slightly changed from the earlier version to clarify the fact that single-sourcing is not *necessarily* the order of the day. In some cases, no one supplier may be able to provide the capacity that you need. In others, while the capacity is there, it may be that none of the potential suppliers is good enough to be trusted with the responsibility and privilege of becoming your single supplier. Going for a single supplier may be something of a gamble, one in which both the potential gain (if successful) and loss (if unsuccessful) are great. Should you enter the gamble unless the chance of success is good? Will it reduce variation? For that is the purpose.

There are many other arguments often raised in opposition to single-sourcing. Deming has nipped some of these in the bud pretty sharply. For example:

"If you feel that a single supplier is likely to produce complacency, laziness, corruption, falling behind, then the foundations have not been laid."

To a question from my colleague David Lobley, then of NEDO (the National Economic Development Office) and now of the Cabinet Office: "Isn't competitive tender a way to avoid corruption?" Deming's retort was:

"Competitive tender fosters corruption."

"What about protection against emergencies?" asked another delegate. In answer to that, Deming started to speak in terms of customer-supplier *partnership*. He is not calling for reduction to a single supplier as some dogmatic principle. And the mere act of awarding business to a single supplier will not achieve miracles in isolation—indeed it can, for obvious reasons, be hazardous. A further delegate raised the possibility of another supplier coming up with important innovation. Deming pointed out that therefore you had chosen the wrong supplier, or that the two of you had not been using the relationship to best advantage. It must be a relationship of *trust* and *action*. A customer-supplier partnership must be dedicated to continual improvement. "What about protection against fraud?" asked yet another delegate:

"Is that the kind of partner you choose? Who's guilty?"

So why even think of aiming for a single supplier? The short, but very important, answer brings us right back to the core of the Deming philosophy: to *reduce variation*, the fundamental requirement for improving quality. And we can go only so far with reduction of variation when we have multiple suppliers. However good they are, and even if they could be regarded as of equal quality with each other in some respects, multiple suppliers are, of course, all different: they have different locations, different systems, different processes, different people, different sub-suppliers, and so on. All these differences are extra sources of variation compared with the case of a single supplier. Multiple suppliers are therefore a bar to continual improvement. Deming became acutely aware of the need to decrease variability in the quality of incoming materials during his first visit to Japan after the war:

> "Incoming materials were terrible, off gauge and
> off colour, nothing right."

(*The Keys to Excellence*, page 20). But:

> ### *"Somehow they made it work—*
> ### *maybe only the Japanese could have."*

Further, whereas the variation from one supplier can be bad enough, the additional variability caused by swapping between suppliers is self-inflicted injury. People on the shop floor know well that adjustments of machinery to compensate for such additional variation is time-consuming and costly (*Out of the Crisis*, page 35). Deming quotes one sufferer as follows:

> "We can learn to use just about anything,
> as long as you keep it roughly the same."

He recalls one seminar where he had eight people up on stage talking about the problems they had had with single suppliers:

> ### *"But none of them would change back to two."*

No wonder, for:

> ### *"I have yet to see a customer*
> ### *who has enough knowledge*
> ### *to work with one supplier.*
> ### *Don't tell me he has enough to work with two."*

Returning to the emergencies referred to in an earlier question, Deming observes that two suppliers will give you twice as many fires, twice as many strikes, twice as many of your vendors going out of business. Indeed, the emergencies alluded to by the questioner—fires, strikes, accidents, disasters of any kind— are *more* likely under traditional arms-length relationships; in particular, if business tends to be awarded on the lowest price, the very cost-cutting needed to win contracts leaves less opportunity to prevent those emergencies. Also, if a supplier is running into some kind of trouble, the last thing he will do in the traditional environment is to let his customer know about it. In the partnership culture, that will be one of the earliest steps. Maybe the customer can help; or, if not, at least he will have been given fair warning.

Whatever the pros and cons, there is no doubt that major companies have made considerable moves toward the reduction of numbers of suppliers. Figures are available to indicate that the motor manufacturers have reduced their number of suppliers to something like 10% of what they were a few years ago. Deming mentions Northern Telecom's reduction from 211 suppliers to 15 over a five-year period. A delegate asked: "What happened to the other 196?" Not all of them went to the wall. Some became sub-suppliers to the 15, while Northern Telecom helped others to find new business. Not all of the 196 were bad companies: there was no value in destroying them.

The partnership culture is a prime ingredient of a single-sourcing policy. The purpose must be to enable the construction of a genuine, long-term, mutually beneficial relationship, involving trust and friendship. How long is "long-term"? "Why limit it?" asks Deming. We are talking of a relationship where a genuine handshake is more powerful than a legal contract. This has not been unknown in the past. It's not unknown now. Deming has reported, in particular, that the Navistar company is now engaging in no written contracts with its suppliers, business being confirmed by handshake, with legal obligation replaced by honour. More of the same is needed.

Partnership is all about helping each other. Suppliers and customers *need* help from each other. Deming sometimes introduces World Number 4, in which the customer thinks he knows what he needs—but he doesn't really. That's getting even closer to the real world. A supplier often has lots of relevant knowledge that the customer cannot possibly have.

The reverse is also often true, e.g. see pages 196–197 in *Out of the Crisis*. Deming's well-known antipathy to the "dishonest" coathangers found in most hotels, and his insistence that hotel managers get to hear about it (see Chapter 23), might be included in this reverse category, as indeed is true of all relevant feedback. The question is often raised, after seeing the failure of all the conventional management efforts in the Experiment on Red Beads, as to how the situation could be improved. The obvious answer is to work with the supplier to reduce the number of red beads in the raw material: maybe the supplier doesn't even know that they are harmful or useless.

At one seminar Deming suggested that, rather than people just staying with their colleagues in the working groups, they should instead work with their suppliers and customers, pointing out that "incest might be the wrong thing"! A customer-supplier partnership-relationship is an obvious application of the Cooperation: Win-Win philosophy. That relationship is not feasible under a policy of multiple-sourcing. The relationship is for each other's benefit. Its spirit is encapsulated in the words of Robert Brown of the Nashua Corporation:

> "This is what I can do for you.
> Here is what you might do for me."

(*Out of the Crisis*, page 43).

Once entered into, it is vital to the interests of both partners that the long-term relationship works. James Sherman, Purchasing Manager for Kimberly-Clark, points out the supreme importance to the customer-company that their suppliers do not go out of business, for then they *both* lose. The policy is Win-Win, not Lose-Lose.

The security of a long-lasting relationship better enables the supplier to innovate. How can a customer expect a supplier to invest anything very solid into a short-term contract—indeed, how can a supplier *afford* to? In the new economic age, the boot will move to the other foot—suppliers will need to choose their customers with care:

> *"Companies write to me*
> *for references on my students.*
> *I tell them:*
> *'Don't worry about my students.*
> *Tell me what are your qualifications*
> *for seeking them.' "*

Already, it is not entirely unknown for a supplier to quit because his customer will not work with him. Ron Moen (whom many will know from the video[2] on the Pontiac Fiero project, and who is also a member of the API group) cites the case of a supplier working on a four-year development project. Three years and four months went by before the customer learned that the supplier was unable to succeed. Where was the relationship, the partnership, to allow this to happen?

Joyce Orsini, already mentioned in the previous chapter, tells of a supplier-company doing its best to serve a large customer. But the customer kept changing his plans, and the supplier lost $2,000,000 over five years through making product that this customer ordered but eventually decided it didn't want. A customer has a moral obligation, if not a legal one, *not* to treat sup-

---

[2] *Roadmap for Change—The Deming Approach*

pliers in this way. Eventually, the time came when the customer sent through another large order, but the supplier didn't respond. The customer made enquiries. The answer came back: "We have regrettably concluded that it is too costly to do business with you." Fortunately, the customer listened, and the incident led to a much more cooperative relationship between the two companies, to their mutual benefit.

It is, of course, normally the customer who has to take the initiative. Potential suppliers can advertise and make themselves known in other ways, but it is the customer who has to consider purchasing policy, in particular the decision whether or not to go for single-sourcing.

How should a customer set about choosing a supplier? With great care! This is not something to jump into, especially if the single-sourcing principle has been accepted. There is much to discover. Get talking to the possible suppliers. Find out how they are doing on the 14 Points. Find out about their processes and systems, and their approach to statistical control and improvement. Where is their knowledge, and what is it? How do they stand with respect to the Four Prongs of Quality (Chapter 14)? Get a look at their financial reports. Are they investing in the right things? Is money being spent on training on the job (Chapter 24), and on education (Chapter 31), and on the pension fund? Won't they tell you, or let you look? Then where's the chance of establishing that long-term trusting partnership? What have they done for you in the past? What have they done for your competitors? What about capacity? Would you be, say, just 5% of their business? Would you want to be? Would you be important to them?

The guiding principle for the choice must again be that of Cooperation: Win-Win. The supplier must be enthusiastic to develop specialist-knowledge about the needs of the customer beyond those which *either* of them currently understand, in order to improve product and services. The problem of choosing a single supplier has given rise to one of Deming's strongest statements:

> *"The overriding requirement for a single supplier*
> *is his* **burning desire** *and ability*
> *to work with you on a long-term basis."*

It is this which you must take the greatest care to assess. It surely cannot be done by standard methods of supplier-assessment. You will need to get to know the people.

Chapter 23

POINT 5: IMPROVE EVERY PROCESS

Improve constantly and forever the system of planning, production, and service, in order to improve every process and activity in the company, to improve quality and productivity, and thus to constantly decrease costs. Institute innovation of product, service, and process. It is management's job to work continually on the system (design, incoming supplies, maintenance, improvement of equipment, supervision, training, retraining, etc.).

Much that has already appeared in this book bears direct relevance to Point 5. What is improving "constantly and forever the system" but the "Obsession with Quality" at the head of the Joiner Triangle? What is the Deming Cycle but a plan to aid continual improvement and innovation? In Chapter 21, we talked about the need to eliminate mass inspection as the *way of life* to achieve quality. What is a better way of life? Surely, continual improvement and innovation. The Taguchi loss function (see Chapters 11 and 12) enters the picture in two guises—first, as an argument for continual improvement rather than the mere meeting of specifications (which is not continual improvement but improvement only up to a standard), and secondly as a help in prioritisation of the candidate-processes, realising that we cannot do everything at once. The reference to innovation in this Point (a relatively late addition) brings us back to the Four Prongs of Quality (see Chapter 14), and the realisation that the improvement of processes, though necessary, is not enough for survival and success.

What must we watch out for in the way of *mere delusions* of improvement? Let us recall that tampering (treating common cause variation as if it were special cause) is *not* improvement of the process. It is costly. As we have seen, sometimes it is literally better if people do *not* put forth their best efforts; if they do not know what to do or why, it is often better to leave well (or even ill) alone.

We have learned in Chapter 4 that putting out fires (tackling special causes) is not improvement of processes. Firefighting, patchwork, cosmetics, solving problems are all to the good when necessary. The point is that they should not be necessary. They don't even do the job of improvement, let alone innovation. Their necessity is a bar to both. There is so much to learn. Thus, both training and education (Points 6 and 13) are vitally important.

Deming provides many anecdotes of failure to improve, and of needless deterioration. Considering the vast amount of travelling that Deming undertakes, it is hardly surprising that many of his stories concern hotels or aeroplanes. There are several in *Out of the Crisis*, especially on pages 95–96 and 212–214. And, as he asks on page 51, why is every hotel completed now not better than ones completed two years ago? Can we not learn? Are we not interested in learning?—

"Most hotels are built for the management without thought for the customer."

A familiar story concerns coathangers in hotels. Deming abhors the fact that so few hotels nowadays provide "honest" coathangers. It *is* irritating and petty, isn't it? He lists on page 213 of *Out of the Crisis* a few that do. Do they charge more than other hotels in order to defray the losses? He is keeping up the campaign. A few months ago it was apparently the turn of the Sheraton Airport Hotel in Portland, Oregon. Deming returned his room-questionnaire, ignoring all the printed questions. But he had added one of his own:

> "Question 49: How do you like our coathangers?"

and also his answer:

> "I don't!"

He reports that he subsequently received a letter from the hotel manager. It read:

> "Thank you for being our guest, and for letting me know that you don't like our hangers. I don't either, and we will be changing them over to regular coathangers within the next two months."

It may be noticed that all the examples cited in this chapter are to do with service rather than manufacturing operations. Why not? One of the clear indicators of lack of understanding of Deming's work is the implication which we sometimes hear that, since it's statistical, it must all be to do with mass-production processes. Quoting again from the Preface to *Out of the Crisis*:

> "The book makes no distinction between manufacturing and service industries ... "

Nor does Point 5. For:

> " ... All industries, manufacturing and service, are
> subject to the same principles of management."

However, to be helpful, the very substantial Chapter 7 in *Out of the Crisis* is dedicated to illustrations and applications of the principles in service organisations.

There is, of course, a huge range of service organisations. Deming gives a long partial list on page 184 of *Out of the Crisis*, and wishes he had also remembered to include leasing on that list. Figures from pages 184–185 are worth reproducing: 75% of employed Americans are engaged in service industries.[1] And nearly half the remainder, though employed by manufacturing organisations, are involved in the service side of those organisations. That leaves only 14 out of 100 "to make items that we can drive, use, misuse, drop, or break." One particular figure indicates very clearly how the world has changed. The 14 out of 100 includes agriculture; but this is now less than 2% of the total, whereas it used to be 50%.

Although the decline of manufacturing in our countries is to be deplored, poor quality in service industries and service operations, like administration, order-processing etc., contributes to the negative balance of trade (see Chapter 1). Many organisations suffer from huge numbers of errors in paperwork—one instance that Deming mentions involved 40% of invoices containing errors. Think what that means for inefficiency, poor reputation—and accounts receivable.

I have never understood why so many people think that quality improvement is more difficult in service than in manufac-

---

[1] Richard Kay has provided me with a press–cutting (*Sunday Times*, London, 10 December 1989) on the American unemployment situation. Part of it reads: "All of the November increase in employment was in the service sector. This was seized upon by economists as a sign of weakness of manufacturing ... "

turing operations. Perhaps that is because most of my limited experience has been on the nonmanufacturing side. From what I have seen, data in service and administration can be easy to come by, processes (both actual and intended) are often much easier to investigate and flow-chart, and the operations involved are often relatively non-technical and thus capable of being understood by others than the "experts." Deming seems to agree:

> *"Quality of service is* **so easy** *to measure, and you usually have the result straightaway. In manufacturing, it can take years."*

Banking is, of course, a service: a vitally important one. Deming reports an occasion on which the Vice-President of a bank attended his seminar and, presumably without looking in the index of *Out of the Crisis*, enquired of Deming whether the book contained any specific advice about banking. Deming turned to his index, and found about a dozen references, including a long section in the Chapter 7 which we have just mentioned. This also gave him the chance to point out a paragraph in that chapter which makes clear the crucial role that banking could play in the changes that need to come about:

> "Banks, by focusing on long-term capital gain, seeking loans to companies that have adopted the 14 Points ... , instead of focusing on short-term results, could help American industry, as banks in Japan help Japanese industry."

As he pointedly remarked:

> *"Quality in banking might be worth a thought."*

Chapter 24

POINT 6: INSTITUTE TRAINING

Institute modern methods of training for everybody's job, including management, to make better use of every employee. New skills are required to keep up with changes in materials, methods, product design, machinery, techniques, and service.

Training is for skills (unlike education, which is for the development of knowledge—see Chapter 31). It is for learning how to do a particular job in a particular way. How can a worker carry out a job if he does not know what the job is?

To be this specific about the purpose of training has important consequences. First, because of the clear focus, it really should be possible to develop training which is maximally effective and leaves little if any room for doubt in the mind of the trainer about what is required. This is an important contribution to the reduction of variation which, as we know, is essential for the improvement of quality. Second, it leads to unexpected and beneficial theory on the value of training and retraining as it relates to a specific job, a specific training programme, and a specific trainee.

These two properties of training mark it out as being totally different from education. First, education cannot be that specific; education is development, growth, expansion; it has no bounds. There can therefore be no categorical definition of what we need to do in furtherance of education. Similarly, there can be no thought of defining when an educational task is completed. Who would know? How would they know? Often the educator learns more than those whom he strives to educate. And the more he strives, the more he learns.

Incidentally, a trainer needs good understanding of operational definitions (see Chapter 7). The purpose of training is to teach skills for a job. Unless the trainer understands that job in an unambiguous, clear-cut way, how can he train others for it? Successful execution of the job will involve properties such as clean, satisfactory, careful, correct, attached, tested, level, secure, complete, uniform, consistent, balanced, vertical, dry, smooth, equal. Without the trainer understanding and, in effect, providing operational definitions of such relevant characteristics within the training procedure, the trainees will depart with different beliefs, and that particular job will not get done in a particular way.

Two main aspects of training on which Deming concentrates are that:

1. People learn in different ways; and
2. Once a worker has brought his work into statistical control, further lessons will not help him.

We shall consider these in turn.

The first aspect is familiar to us as Item 3 of Part D of the System of Profound Knowledge (Chapter 18). Many learn best from illustrations and pictures, and others from demonstrations; some prefer the written word to the spoken, others *vice versa*. Many people have been cashiered out of the army because of not obeying instructions they couldn't understand (*Out of the Crisis*, page 52). Years ago, Deming was puzzled by the lack of response when speaking to the famous statistician Harold Hotelling. His friend W. Allen Wallis (previously mentioned in Chapter 2) told him that Hotelling had difficulty in understanding the spoken word. Thereafter, Deming corresponded with Hotelling in writing, and the communication then became two-way.

A tricky concept, which deserves more debate and more application than it gets, is Deming's contention that training is a one-chance opportunity. If we get it wrong, the damage is permanent. This topic is well-covered in the early pages of Chapter 8 in *Out of the Crisis*. However, we will outline the main points here. To briefly quote from *Out of the Crisis*, page 249:

> "Anyone, when he has brought his work to a state of statistical control, whether he was trained well or badly, is in a rut. He has completed his learning of that particular job. It is not economical to try to provide further training of the same kind."

What is implied by "brought his work to a state of statistical control?" Surely, that the stage has been reached where the worker's performance in this skill has become *predictable* (in the usual sense of Chapter 4). That is quite different from his performance still being out of statistical control, i.e. unpredictable, because he is still unclear what the job is. In that case, further training will help. In the former situation, it won't: in that case, he knows what the job is, or *believes that he does*. If what he believes is wrong, or if it's not exactly wrong but it's still not good enough, that is the fault of the training. It is not he that has failed: it is the system of training which has failed.

If that is the case, there are now three options to consider. Option 1 is simply to cut the losses, admit that the system has failed, transfer the worker to a different job, and try to be more successful in training him for that. Sure, that sounds drastic and expensive. But is it more expensive than leaving him on the *same* job with the almost sure guarantee that he will carry on doing that job badly (remember, we have agreed above that his performance has become predictable—predictably bad!—due to the unsuccessful training)? It is little excuse to protest that perhaps, right from the start, he was incapable of doing the job:

> ## *"If a worker cannot learn his job,*
> ## *why did you put him there?"*

Is it not more a case of you, as manager or recruitment officer, failing to do *your* job?

The other two options are:

2. To put him through the same training again; or
3. To try him on some different training for the same skill.

The theory of common and special cause variation indicates that Option 2 would be fruitless. The stable performance tells us that only common cause variation is present—i.e. that the *minimum* variation possible under the current system has been reached. Only change to the system itself has a chance of improving matters, and sending him through the *same* training course again is hardly change to the system. Quoting again from page 249 in *Out of the Crisis*:

> "When a person has reached the state of statistical control ... , continuation of training *by the same method* will accomplish nothing."

(my italics).

Option 3 is more arguable. We are not now talking of training *by the same method.* Training by another method may be successful. Deming gives a case history. A 30-year-old woman had been trained for a job. Her performance in that job had stabilised, i.e. had become predictable, but was not satisfactory. Her manager moved her onto a different job for the best part of a year while he worked on new training for the old job. A year later he brought her back, put her through the new training, and her work in that job eventually became entirely satisfactory.

The two lessons to learn from this story are that training by a new method for the same job *can* work, but it is liable to be *expensive.* Much better to get it right the first time (to coin a phrase!). However, it is clear that Deming regards even Option 3 as generally undesirable—he believes that the expense and difficulty will not usually be justified. Why? I think the basic reason is the well-known human characteristic that bad habits are much harder to dispel than no habits! Once something has been learned wrong, learning it right consists of two parts: getting rid of the wrong and then receiving the right. The former is *difficult*, very difficult.

You must have seen examples. Perhaps you taught yourself to type by the "hunt and peck" or "two-finger" method. If so, did you ever subsequently learn touch-typing? If you were brought up with bad eating habits, have you ever really changed to a healthy diet? If you learned to play the organ without good pedal technique, do you *still* look at your feet except in the easiest of passages? If you were taught to speak French by someone who didn't, how confident do you feel now, 10, 20, 30 years later, when trying to converse with your hotelier, restaurateur, or shopkeeper in Paris?[1] I am not saying that reform in these or in

---

[1] Teaching a language is susceptible to Rule 4 of the Funnel (Chapter 5). If one learns the grammar and vocabulary, but not the accent, and then teaches another, who then teaches someone else ... , what will be the result? This again illustrates the hazard of "worker training worker."

countless other cases is impossible: but I am saying that it is darned difficult. Old habits die hard. And those habits, one way or another, were *trained* into you, and your performance in them did become predictable. Years later, the chances are that that performance hasn't changed much.

A number of examples are described in the early pages of Chapter 8 in *Out of the Crisis*. Two of them involve golfers. One is the case of a golfer whose performance was out of statistical control but who then clearly benefited from lessons, and the other concerns a golfer whose performance was in statistical control and who then had lessons which turned out to be entirely ineffective. These examples were used by Deming in his first courses to the Japanese in 1950. Incidentally, his own experience with the game doesn't seem to have been overly successful:

> *"Golfers are an unhappy lot.*
> *I played a game myself once—in 1929.*
> *Lost the ball, and quit!"*

The case of language teaching mentioned above is discussed at greater length in *Out of the Crisis*, pages 254–255. Again, the basic difficulty is that of getting rid of bad habits. A further example of the problem of getting rid of bad habits, as opposed to just acquiring good ones starting with a clean slate, comes from Heero Hacquebord. Heero told me of how he had been badly trained in tennis at school. He had recently tried to take up the game again; but the bad training had stuck. Eventually he gave up tennis as a bad job and took up golf—in which, incidentally, he ensured he received some *good* training (presumably unlike Deming!).

Before we leave this issue, let us revisit the case of the Western Roll and the Fosbury Flop (Chapter 20). Perhaps the analogy of the two approaches to the high-jump with bad and good management should not be pursued too far. For it appears that no athlete who had achieved excellence using the Western Roll managed to convert very successfully to the Fosbury method. I suppose we should conclude that high-jumping is a skill, thus requiring training, whereas management is knowledge, thus requiring education. In either case, unlearning is always more difficult than learning. Management surely needs to engage in the unlearning and relearning as "All One Team," as it is too large a task for individuals. And will widespread understanding have to wait for another generation, a generation with less unlearning to carry out? If so, the outlook is bleak indeed.[2]

Training is often one of the first things to suffer if finances get at all tight. Brian Joiner has pointed out (*Out of the Crisis*, page 53) that money spent on training, unlike money spent on equipment, does not contribute to the tangible net worth of a company under traditional accounting practice. In contrast, a financial constraint sometimes imposed on Japanese managers is the large *minimum* amount which they have to spend on training.

A final thought comes from a manager who told Deming: "We do little training here, because of the high turnover of staff." Could you think of a better example of a self-fulfilling prophecy?

---

[2] Alistair Morrison, Managing Director of Skillchange, has elaborated usefully on this paragraph as follows: "In most cases, both education and training are required for success. Thus the high-jumper must be *educated* as to the features and benefits of the Fosbury method prior to *training* in how to actually do it. Equally so for managers! Management is not just knowledge; it is also *application* of knowledge. Managers must be *educated* to gain the knowledge, and also *trained* in its application." Alistair also points out that, especially with the rapidly declining birth-rate in the West, the need for both education and training to figure clearly in the minds and actions of senior management is, if possible, even more essential.

Chapter 25

POINT 7: INSTITUTE LEADERSHIP [1]
OF PEOPLE

Adopt and institute leadership aimed at helping people to do a better job. The responsibility of managers and supervisors must be changed from sheer numbers to quality. Improvement of quality will automatically improve productivity. Management must ensure that immediate action is taken on reports of inherited defects, maintenance requirements, poor tools, fuzzy operational definitions, and all conditions detrimental to quality.

Point 7 has assumed an even greater significance since Deming changed "Institute supervision" to "Institute leadership"

---

[1] During 1991, Deming began to reserve the word "leadership" for what we have termed "leadership of transformation" in Chapter 18, now referring to the concepts in this chapter as "management of people." I believe the concepts here to be so important that I prefer to retain the stronger term "leadership" or, more fully, "leadership of people."

in 1986. Indeed, the genuine adoption of Point 7 is fundamental in Deming's answers to the usual questions which arise out of the controversial Points 11 and 12, such as "How do you motivate people if you don't give them (arbitrary numerical) goals?" and "How can you ensure they do a decent job if you don't carry out appraisals of their performance?" The response "Adopt leadership" might sound rather lame until we start learning what Deming means by that word "leadership." (Recall also the reference to "leadership of people" in the Summary of the System of Profound Knowledge near the end of Chapter 18.) He moved away from the word "supervision" because his thoughts had nothing to do with common connotations of supervision (i.e. chasing quotas and mistakes, time-keeping, etc.).

Incidentally, the words "supervisors" and "supervision" still appear here and there in the text of *Out of the Crisis*—that is intentional: they do refer to the positions and tasks normally indicated by those words. However, the issue being clarified here is that this is *not* what he is requiring us to "institute" in Point 7. Indeed, supervisors themselves need to develop the leadership abilities discussed in this chapter. There was a sign which Deming saw at Ford, put up in jest by a plant manager:

"The floggings will continue until morale improves."

He used to think that was funny; now he's not so sure.

The function of leadership, in the words of Point 7, is to help people do a better job, a natural consequence of which will be to improve the product, service, and profits of the company. At BOC Powertrain, the Finance Director's job has been newly defined by President Bob Stempel not as to cut costs but to enable people to do a good job.

A page headed "Some Attributes of a Leader" has been circulated at several seminars. Figure 50 is a reproduction from the original version of that page. As is so often the case with Deming, a few words imply a very great deal.

SOME ATTRIBUTES OF A LEADER

1. Understands how the work of his group fits in to the aims of the company.

2. Works with preceding stages and with following stages.

3. He tries to create for everybody joy in work. He tries to optimize the education, skills, and abilities of everyone, and helps everyone to improve.

4. Is coach and counsel, not a judge.

5. Uses figures to help him to understand his people and himself. He understands variation. He uses statistical calculation to learn who if anybody is outside the system, in need of special help.   Transfer to another job may require prudence and depth of understanding. The man transferred may take it as one way to get rid of him.

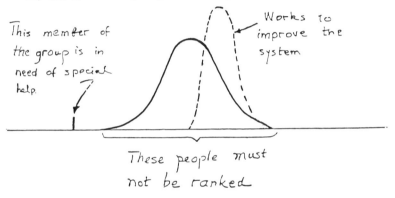

This member of the group is in need of special help.

Works to improve the system

These people must not be ranked

6. Works to improve the system that he and his people work in.

7. Creates trust. He is aware that creation of trust requires that he take a risk (Carlisle and Parker).[2]

8. Does not expect perfection.

9. Listens and learns without passing judgment on him that he listens to.

FIGURE 50
Some Attributes of a Leader
(Prepared by Dr. Deming)

---

[2] Attribute 7 refers to an instructive game described in *Beyond Negotiation* by the cited British authors. It teaches the need for courage above the interests of normal self-preservation in order to improve circumstances for all.

A careful study of the list in Figure 50 will provide some understanding of why "Adopt leadership" *is* a very substantial answer to the questions posed above concerning Points 11 and 12. Incidentally, an expanded list shown to me by Bill Scherkenbach also includes: "Improves continually his education" and "Creates more leaders." Further related material may be found in *Out of the Crisis*, pages 116–118.

The reader will see that many fundamental features of the Deming philosophy are represented in the list—cooperation, joy in work, the understanding of variation, and improvement of systems. Notice, in particular, the way each attribute involves *people* in an understanding, constructive, sympathetic, and humanitarian way. This is one of the many aspects of this philosophy which make it so different from other approaches to quality improvement:

> *"I used to say that people are assets,*
> *not commodities.*
> *But they are not just assets:*
> *they are jewels."*

A requirement of leadership, implied in Attribute 5, is the ability to understand a system:

> *"A leader must understand*
> *that the system is composed of people,*
> *not mere machinery,*
> *nor activities,*
> *nor organisation charts."*

In particular, it is the ability to analyse special and common causes in respect of the systems within which people work. Fundamental in this is the avoidance of any hint of blame on people for problems (the vast majority) caused by the system, and the aim to help (instead of blame) the few people whose performance does fall below that of the main system:

> *"A leader's job*
> *is to help people, not judge them.*
> *It is to know*
> *when people need special help,*
> *and provide it.*
> *He is not a leader unless he does know."*

Again, of course, this is not new—see, among other examples, pages 195 and 256–261 of *Out of the Crisis*. Note the insistence on avoiding judgment of people; it occurs in both Attributes 4 and 9.

Attributes 5 and 6 may need a little further explanation. The hand-drawn sketch shows, in typically economical fashion, much of what is involved in these two Attributes. Deming's concentration on *variation* has shifted us from the traditional emphasis on *average*. The plea to "keep it roughly the same" in the discussion of Point 4 (Chapter 22) is a simple indication of why this shift is valid. But such a change of emphasis does not, of course, imply that average values are unimportant. The sketch following Attribute 5 shows *both* decreased variation *and* raised process average as representing an improved system. The fact is that reduction of variation often automatically results in the bonus of improvement in average level of performance. Regrettably, more traditional attempts to directly improve averages often result in *increased* rather than decreased variation—as is certainly the case if the goal of an improved average is beyond the capability of the system.

The insistence in the sketch that "These people must not be ranked" is fundamental to the arguments on performance appraisal (see the discussion on Point 12 in Chapter 30). There is no justification for ranking performance of people whom statistical techniques (control charts) fail to indicate as special causes: if the figures cannot be differentiated from the behaviour of mere random variation, what reason can there be for differentiating between the people? In any case:

> *"Improve the system,*
> *and variation between people will diminish."*

Further, even if some statistical signals are seen, the Pygmalion Effect (again see Chapter 30) may well provide the main explanation. Indeed, all discussion of Point 12 is highly relevant to the topic of leadership.

A recent emphasis by Deming, which is implied in a number of the Attributes but does not appear explicitly, is the need for leaders to be proactive in their search for difficulties and in their desire to help those for (and to) whom they are responsible:

> *"People do not tell you their problems;*
> *the supervisor's job is to find out what is wrong."*

When substantial moves have been made in adoption of the Deming philosophy, this emphasis will be less important. When the true spirit of "All One Team" has been developed, and people are genuinely without fear, they will freely share their concerns with those whose job it is to help. What would they have to lose? They could have much to gain. But, before we reach that happy situation, people will hold back.

A leader's (supervisor's, foreman's, manager's) job *now* is to remedy that—and thereby *speed* the progress to the improved situation. A worker's problem may be with machinery, with procedures, with documentation—or it may be personal. Whichever it is, if the leader takes the trouble to find out, then he can both aim to help the person and improve the overall system within which he and his colleagues are working. Belief that the company's management really does care will grow, and Cooperation: Win-Win along with it.

Considerable education and training are surely necessary in order to help create good leadership. Unfortunately, in practice, people are often thrown into positions of responsibility with little or no such education and training. As a delegate at a four-day seminar said: "Our supervisors are college boys and girls who have studied human relations." Closely related to this is Jim Bakken's contribution to *Out of the Crisis*, page 55. Very appropriate also is a quote from Myron Tribus:

> "You can manage what you do not understand,
> but you cannot lead it."

Contrast all this with the example of Mr. J. C. Capt who, according to Deming, when put in charge of the National Bureau of the Census, imposed no changes at all for about a year, i.e. until he had had some chance to learn the work. In so doing, he made sure that such changes as he did eventually make were based on solid information and knowledge, and he also generated a feeling of confidence, worth, and security on the part of everybody in the organisation.

Deming is often asked what qualifications we should look for in candidates for promotion. You could now surely predict his answer:

"What better than ability to be a leader?"

And, in summary:

"Why lead?
People happier, quality up, productivity up,
everybody wins."

Chapter 26

POINT 8: DRIVE OUT FEAR

Encourage effective two-way communication and other means to drive out fear throughout the organisation so that everybody may work effectively and more productively for the company.

Fear is an obstacle to all that needs to be achieved. It is at least inconsistent with, and generally a barrier to, all the main concepts of the Deming philosophy. It is a barrier to improvement and innovation, it is inconsistent with joy in work, it is a barrier to change, it is inconsistent with cooperation, it is a barrier to the Scientific Approach, it sets up barriers rather than breaking them down. It is a weapon in the armoury of traditional management; it is an enemy to be blown up and blown away by the new philosophy of management.

One reason why fear is a barrier to improvement and to the Scientific Approach is that:

"Wherever there is fear, we get the wrong figures."

(see Chapter 10). This recalls an occurrence which emphasised in Deming's mind the serious handicap which fear imposes. He was working with Bill Latzko in a certain company, and it became clear to both of them that the figures they were receiving from the work-force were quite unbelievable. They eventually realised that the work-force were in fear of their management, and had learned long ago to provide figures of which the management might approve, whether or not they reflected the truth.

We have referred in Chapter 21 to the phenomenon of "flinching" in inspection processes, a reflection of the attitude: "When in doubt, pass it." Surely that is directly opposite to what is really needed. In the spirit of Cooperation: Win-Win, we want to be doing the very best possible for the customer, internal or external. In that spirit, something on the borderline would not be "slipped through." It could be unsuitable for the customer, so the attitude should be more inclined to "When in doubt, *fail* it" rather than needing to pass as much as possible in order to avoid possible accusations of unsatisfactory work being directed at you or your colleagues. (After all, isn't conventional management always looking for scapegoats: people to blame?) A borderline result could be a signal that the system of production is in trouble, and so it might need attention straightaway before causing harm to company or customer.

In any case, whenever possible, the aim should be to improve processes so much that, even when anything does start to go wrong, the safety-margin is wide enough to still avoid causing any trouble to customers (we refer again to *A Japanese Control Chart*). But bad management argues against doing much better than keeping to specifications, with the impression that "the better it is, the more it must be costing us." Consequently, process capabilities (see Chapter 12) are constrained to stay rather close to 1; the consequence is that maintenance of production standards is always on a knife-edge, and fear of failure is manifest.

The need, or perceived need, to fiddle figures is obviously

a barrier to improvement. It destroys the raw material for the Scientific Approach. There are many ways in which fear can be reflected in figures. A fascinating example is Figure 23 on page 265 of *Out of the Crisis*. This chart, showing two months' figures of percentage defective per day, indicates an average of 8.8%. However, the points are tightly clustered around the Central Line, exhibiting much less variation than is conceivable from a process which is simply in statistical control with that average.

Deming tells the story on page 266. A belief had spread that, if the proportion defective ever reached 10% on any day, the manager would close down the operation. The inspector couldn't hide the overall average defective, but he could massage the figures (decrease the high ones and correspondingly increase the low ones) to avoid ever admitting that the crucial figure had been reached. In fact, given the sample size used, i.e. 225, the control limits for an 8.8% average defective rate are approximately 3% and 14.5%: so the truth must have been that the 10% figure was often exceeded.

Did the manager *really* make the threat, and if so did he ever mean to keep it? It doesn't really matter whether or not he said it, or meant it. The thing that mattered was that the people and the inspector *thought* he said it and meant it. Not unreasonably, that generated fear. Who could blame the inspector for fiddling the figures? He was trying to protect 300 people's jobs, 300 people who were not to blame for the real figures.[1]

If your people are afraid of you, do you believe the figures they give you? You could, of course, threaten disciplinary action if you discover inaccuracies. Is that good management? You would be getting them to weigh up one evil against another.

---

[1] Some further data from the same study are given as Exercise 10.6 on page 264 of *Understanding Statistical Process Control*, and are discussed in the answer to that exercise on page 335. In this case, a few of the data are recorded as being above 10%; however, they are all below 11%, and consequently the range of the figures is still absurdly small.

How about developing an environment where they are *not* frightened of you, nor of the consequences to them if the figures are not as good as you would like to see, but rather one in which they *want* to give you accurate information because that will help you the most and will enable you to improve matters for everyone concerned?

More examples of fear are given on pages 60 and 267–268 in *Out of the Crisis*. Or are they examples of *anxiety*? In recent years, Deming has been taking care to differentiate between fear and anxiety. Fear is caused by relatively specific problems, whereas sources of anxiety are not directly identifiable. Of the two, anxiety generally has the worse effects. With fear, one knows what one is up against, and can perhaps plan something to fight it. But anxiety has no focus—the cause is neither known nor understood; those suffering from anxiety feel helpless, and can become paralysed to the extent where they cannot act, even if they have some sense of what they ought to do.

Both fear and anxiety are destructive: management exhibiting good leadership will work to reduce them as much as possible so that people have the chance to work well. That is another call to abolish the merit system, for:

"Our system of reward generates fear."

One delegate suggested that some fear or anxiety could be a good thing. Deming thought not. If management feel they need the weapon of fear or anxiety, that is automatic admission of their own failure—yet another case of "making the best of a bad job." Perhaps words are difficult to define here. Maybe, in a sense, a little anxiety is necessary for life. I prefer the word "challenge" instead. Challenge excites us to improve ourselves, to learn. Challenge, yes; but fear and anxiety, no. Challenge is positive, fear and anxiety are negative. Fear or anxiety about failure, for example, causes us to "progress" through the Rules of the Funnel (Chapter 5).

Fear inhibits us from making suggestions for improvement to our work and to the system in which we work. Management might interpret that as troublemaking or, at the very least, criticism. Our suggestion may get ignored. Even if it is investigated, it may turn out not to be a good suggestion after all. Better to keep quiet, to keep out of the way.

And if that is the attitude, so much for joy in work! Joy in work can only come from contributing, the more the better. Fear in work is inconsistent with joy in work. We must be careful of words: is it unknown for some people to enjoy being afraid? Certainly, people enjoy the thrill of the more hair-raising fairground rides; some even enjoy the excitement of war. But what enjoyment, what thrill, can there be in wondering if you are going to be held up to ridicule, if you are going to miss out on the next raise, or if you could possibly be losing your job?

The reader will recall Deming's estimate (Chapter 13) that at most two management people in 100 have joy in their work. What about the other 98?—

"The other 98 are under stress, terrible stress."

What causes stress but fear or anxiety? They are under stress, says Deming, "not from work nor from overwork, but through servitude and a life of knavery in nonproductive work, churning money, battling for or against takeover, or strife to raise the price of the company's stock." Yet again, we see that relatively new emphasis in Deming's teachings, joy in work, being totally consistent with his previous teaching.

One essential for joy in work is trust: two-way trust, so that a worker may feel confident of the management playing fair by him and that he is trusted by them and is deserving of that trust. Deming has recently been suggesting that, as an alternative to "Drive Out Fear," Point 8 could be expressed in terms of "Build

Trust." Without trust there is suspicion, games-playing, suppression and distortion of information, no genuine cooperation. Management must develop trust and become worthy of it themselves.

I have heard Peter Scholtes speak with great effect on this topic. He checks companies' procedures for obtaining permission to take time off work, or for claiming travel expenses, or sometimes even for requisition of stationery. He often finds such procedures inordinately complicated and time-consuming. Many of us know what he means. He then asks management what proportion of their employees would take unfair advantage if the procedures were made more straightforward and omitted some of the checking and double-checking. If the answer is high, isn't that another sign of failure of management? It means either that they have been recruiting untrustworthy people, or alternatively that they have managed (in two senses) to destroy the trust that once was there. But often the answer is low—maybe 5%. Peter then asks them to weigh up the value of the protection which the complicated procedures provide from that 5% against the loss of time, efficiency, and morale that they cause to the 95%. That sometimes inspires some changes!

A fine example often quoted by Peter and his colleagues at Joiner Associates is that of the procedure for obtaining bereavement-leave at the Falk Corporation. The original procedure was defined on two closely-printed pages including, for example:

> "All employees shall receive time off with pay
> up to a maximum of three (3) days of working time
> lost if there is a death in the immediate family.
>
> These days must be within a seven calendar day
> period, the first day of which would be the initial bereavement day paid. However, one of the
> days must be the day of the funeral. If the funeral falls on a non-scheduled work day (Sat., Sun.,
> holiday, during a plant shutdown or during a

period of disability), no loss of pay is involved, therefore bereavement pay will not be made for such days.

• • •

Pay will be for eight (8) hours at the employee's day rate plus average premium for the three (3) months prior to the month in which the time off occurred.

• • •

An employee's immediate family will be considered ... spouse, child, step-child, mother, father, sister, brother, step-parents, grandparents and grandchildren of employee, son-in-law and daughter-in-law, mother, father, sister, brother and grandparents of the spouse. ...

• • •

The personnel department may require verification of death and relationship to the employee."

• • •

After some deliberation, the management scrapped that procedure, and replaced it with one comprised of just 20 words:

"If you require time off due to the death of a friend or family member, make arrangements with your supervisor."

Peter showed me some figures. After the change of policy, the number of absences for bereavement went up by over 20%. So much for that change of heart? No, on closer investigation, the *total* days lost through bereavement-leave were found to have *decreased* by about 36%. The *increase* in numbers of absences signified that some people had previously not taken time off to attend a funeral, either through being intimidated by the original procedure or because the deceased was uncooperative enough not to fulfil the criteria! What a great method for making people feel good about the company! But the decrease in total time lost showed that those who did obtain leave under the original procedure tended to take off more time than they really needed. Wouldn't you? In an environment of management by distrust and conflict, most people will take what they can get. In management guided by the new philosophy, people will take, at most, what they need.

Fear may well be one of the biggest obstacles standing in the way of the needed transformation. The transformation is change—enormous change:

> ## *"We need to change,*
> ## *and we need to know why."*

People fear change. And what they fear, they resist. Or rather, as Scholtes and Hacquebord say in "A Practical Approach To Quality:"

"People don't resist change—they resist *being* changed."

Whichever it is, people fear it. They feel insecure. No wonder. Their destiny could be affected by the changes promised (or threatened), changes which are in the hands of others—perhaps others in whom they have little trust. "Where will change leave me?" Rather the devil you know than the devil you don't.

Maybe the fear is justified. Maybe employees have good reason to be fearful, given their past experience. Why should management be trusted this time? The management have made their own bed, and they will have to remake it very carefully. But suppose the situation is better than that. The fear will still be there; how can it be overcome? A strong recommendation comes in one of the versions of Point 14 (see Chapter 32): "Put everybody in the company to work to accomplish the transformation":

> *"The answer lies, I believe, in a plan by which,*
> *under competent leadership,*
> *everyone will work on the changes required,*
> *with faith that everyone will come out ahead.*
> *Everyone will help to plan his own destiny."*

Or, more briefly:

> *"The need is for everybody to be part of change,*
> *and to belong to it."*

Chapter 27

POINT 9: BREAK DOWN BARRIERS

Break down barriers between departments and staff areas. People in different areas, such as Research, Design, Sales, Administration, and Production, must work in teams to tackle problems that may be encountered with products or service.

This chapter is quite short because there is so much of direct relevance to Point 9 elsewhere in the book. Chapters 8 and 15 are obvious examples, and the same is true of Chapter 30 and, to an even greater extent, Chapter 29.

Reflecting his concentration on Cooperation: Win-Win, Deming has described Point 9 as: "Optimisation over all, Win-Win—instead of suboptimisation, by which each one tries to maximise his own profit." Barriers breed suboptimisation. And the bigger and thicker the barriers, the greater the ignorance of, and lack of concern for, the effects of suboptimisation on other departments or individuals.

Incidentally, Deming points out that one clear symptom

of suboptimisation is proliferation of paperwork, resulting in considerable inefficiency, irritation, and cost. He quotes studies showing that 14% of freight charges are spent on paperwork, the figure in motor freight, in particular, being more like 20%.

The requirement to "Break down barriers between departments and staff areas" is encapsulated in the famous flow-diagram: "Production Viewed as a System," described and discussed in Chapter 8 (see Figure 23). This has been a fundamental component of Deming's teaching in Japan. We have already noted its appearance in the Introduction to the published version of his 1950 series of lectures. In his description of that flow-diagram on page 4 of *Out of the Crisis*, Deming recalls that it was first used at a conference *with top management* in August 1950.

It is top management who, through their style and methods of management, erect the barriers; it has to be they who demolish them. Unfortunately, unlike the way that structures of brick or concrete take a long while to build but can be rapidly brought down with explosive, barriers can be built by management with speed, but can only be dismantled with care and patience—they cannot just be blown up. As Myron Tribus pointed out when recently visiting the BDA, a hypothesis or idea cannot just be destroyed: it must, in practice, be replaced by something else.

In addition to Cooperation: Win-Win and joy in work, we have seen in Chapter 14 that one of Deming's main current emphases is the importance of innovation, as opposed to only improvement. Barriers, and lack of communication, genuine interest, and concern between departments dilute whatever innovatory power still survives. Naturally, innovation impacts in different ways on many departments: Finance, Research and Development, Design, Purchasing, Manufacturing, Marketing, and others. Development of an innovation tends to be sequential: designing, tolerancing, drafting, redesigning, etc., etc.. When the barriers are up, opinions and attitudes in any department are most influenced by how it all affects *them*—more work, more cost, more risk—rather than focusing on the company as a whole and the customer. This, combined with lack of communication, produces results akin to Rule 4 of the Funnel—or even worse than that, since lack of cooperation can impose bias *away* from the target! Far better if all contribute positively and together to turn the innovation into benefit for company and customer alike.

Deming says quite bluntly that, if the boss of every staff area perceives (or indeed is told) that his job is to maximise his department's profits, then the company will fail. One of the most crucial differences between traditional and Deming-style management is seen in the way that shop-floor employees, supervisors, departments, middle managers, and senior managers regard their jobs. Is it to suboptimise, or is it to work for the company? Is it to serve the company and its customers as well as possible, or is it to "look after Number 1"? It cannot be both.

And who decides? It is too easy to say that the employee himself decides. In this, as in all else, the system in which he finds himself has the major say. In what direction does the system of management guide him? Of course, the system of management includes the system of reward:

> ### *"Can you blame someone for maximising his own profit if he gets rated that way?"*

Payment of salesmen by performance and by incentives is an obvious example, discussed at some length in Chapter 29. What and whom is then the salesman's prime concern: his company, or the customer, or himself? It would be good if all three could be satisfied simultaneously. That is not possible in the old way, whereas it is a necessity and natural consequence of the new way.

One of the biggest barriers is often that between the Finance Department and the rest of the company. Deming is concerned at the negative influence that Finance Departments often have, whereas they could be playing such a positive role:

> *"Too often,*
> *the financial people in a company*
> *merely beat down costs,*
> *on the thought that any cost is too high.*
> *They could make genuine contributions*
> *to our economy*
> *by learning the new philosophy,*
> *and by joining in to help*
> *to accomplish the changes that must take place."*

(See also the reference to Bob Stempel in Chapter 25.)

An obsession with cutting costs leads to the same mistakes in the context of internal suppliers as Point 4 warns us against in the context of external suppliers. Deming tells of a lady who wanted to spend an hour with him. She was going to fly from Chicago to New York on a day when he would be teaching there. He could see no problem in finding an hour to spend with her. But he could see a problem with her flight-plan. She intended to arrive in New York at 7:00 am, thus effectively losing a night's sleep. Deming suggested she take a later flight. But she told him that her company insisted she take this flight because, due to a special deal they had negotiated, they would save $138 compared with the normal fare. What, asks Deming, would be the *loss* entailed by this saving? What would be the loss incurred by her inability to work during the meeting that she was to attend? "She would have to prop her eyelids open with matchsticks in order to stay awake."

Deming refers to another company in which the names of those employees who fail to use the lowest fares available are published on a list which is circulated round the company. As he asks, somewhat incredulously: "That's management?"

"Break Down Barriers between Departments." Why? Refer again to the "Example: Huge Financial Advantages of Co-operation" at the end of Chapter 15. That illustration answers the question more eloquently than further words can do here.

Chapter 28

POINT 10: ELIMINATE EXHORTATIONS

Eliminate the use of slogans, posters and exhortations for the work-force, demanding Zero Defects and new levels of productivity, without providing methods. Such exhortations only create adversarial relationships; the bulk of the causes of low quality and low productivity belong to the system, and thus lie beyond the power of the work-force.

Here is a Point that critics are quick to scorn. What can be wrong with slogans? What's the harm in a few posters? Indeed, posters must be able to give useful and helpful information. Surely, slogans are useful to jog the memory on important matters, and thus to aid greater commonality of thought and aims.

Of course, that may well be true. So what is Deming on about here? Let's do what the critics do not do, and examine the words and the principles more closely. The very first sentence of explanation on this Point (*Out of the Crisis*, page 65) says:

> "Eliminate targets, slogans, exhortations, posters
> for the work-force *that urge them to increase pro-*
> *ductivity.*"

(my italics). I have included in the above version of Point 10
three words from earlier versions: " ... without providing methods." Deming's basic argument is directed at those who simply
tell (exhort) others to do better without helping them so to do. It
demonstrates ignorance or disbelief of the fact that the large
majority of problems lie in the system, the responsibility of management, rather than with the workers at whom the posters and
slogans are aimed.

Worse still, posters are sometimes used as a blatant abro-
gation of management responsibility. The display of a poster
attempts (and often succeeds in the eyes of the law) to shift
responsibility for management's shortcomings to the individual.
Heero Hacquebord tells the story (*Out of the Crisis*, page 316) of
how he saw the prominent sign: "YOUR SAFETY IS UP TO
YOU," and then nearly fell off some steps to his death because
those steps were so rickety. Further, I always feel some unease
when I drive my car into a public or hotel car-park and pass the
notice which emphasises that the proprietors accept no responsibility for theft, damage, etc.. Why not? I'm paying for the
privilege to leave my car in their park. What am I supposed to
do? Stand and watch my car all night?

Deming considers that people are already doing their
best. Even if you don't believe that, do you expect them to suddenly start doing their best because of such management gimmicks? The posters and slogans being referred to in this Point
anger those who are already contributing all they can; and they
will be ignored by any who are not. Such exhortations as "Take
pride in your work" are *degrading* to those prevented from doing
so by the system that they cannot influence. The only possible
improvements that can be affected by slogans and posters are in
the minority realm of *special* causes—as Deming himself says in

Out of the Crisis, page 67:

> "The immediate effect of a campaign of posters,
> exhortations, and pledges may well be some
> fleeting improvement of quality and produc-
> tivity, the effect of elimination of some obvious
> special causes."

Sounds OK? I suggest you go to his book and read further on that
page! For, in effect, some approaches to quality improvement
concentrate unduly on special causes and far too little on common
causes.

Exhortations such as "Take pride in your work" have no
purpose in a Deming environment. In that environment, people
are already doing their best. What they need in order to improve is help, advice, training, and a better system within which
to work—in other words: *leadership of people*, just as described in
Chapter 25. How about "Do it right the first time"? What good
is it for the foreman to say that in the Experiment on Red Beads
(Chapter 6)? If only they had the chance, instead of having to
deal with products and systems which are already defective.
The exhortation (as in the Experiment on Red Beads) may even
include a numerical goal, but we shall reserve discussion of that
until the next chapter. Some fruits of such exhortations are listed
on page 68 of *Out of the Crisis*. It could also be a useful exercise to
read through the extract from a Navy yard bulletin (*Out of the
Crisis*, pages 69–70) to see how many "horrible examples" can be
found!

But what about those "posters" and "slogans"? Are they
really all excluded, according to Point 10? No, of course not.
Posters and slogans whose genuine purpose is to help, advise, and
communicate are fine. Information in poster-form indicating how
management's genuine constancy of purpose is being developed
and applied would be good. See also the paragraph beginning at
the bottom of page 68 of *Out of the Crisis*, ending sadly with "I

have not yet seen any such posters." A useful example (which *does* exist) is given in *Out of the Crisis*, page 193. Further, posters designed by workers themselves for their own assistance are highly likely to be beneficial.

I sometimes think that people's reaction to Point 10 can be a useful guide on how far they have or have not progressed in appreciation of Deming's work. It is not a difficult Point. I fear that those who jibe at it, as at the beginning of this chapter, are still near the starting-post; for example, they cannot yet have come to grips with the elements of Shewhart's original work and its ramifications as discussed in Chapter 4.

Chapter 29

POINT 11: ELIMINATE ARBITRARY NUMERICAL TARGETS

Eliminate work standards that prescribe quotas for the work-force and numerical goals for people in management. Substitute aids and helpful leadership in order to achieve continual improvement of quality and productivity.

At first sight, Point 11 is one of departure between Deming and the Japanese. It is well-known that Japanese companies often use detailed numerical goals. A Japanese manager was asked what happens if the goals are not met. His answer was instructive:

"Then management must analyse the system"

—must analyse the *system* to see why *it* had failed to live up to expectations, rather than seeking to blame the people working within that system.

No, Deming is *not* telling us to manage without numbers. Of course, both companies and individuals need goals, intentions, aims, objectives. What they do *not* need is *arbitrary numerical* goals. Deming succinctly expresses the difference on page 90 of *Out of the Crisis*:

> "A group, a team, should have an aim, a job, a goal. A statement thereof *must not be specific in detail*, else it stifle initiative."

(my italics). Of course, a company needs budgets and forecasts for planning and allocation of resources—but *they* must also not be arbitrary numerical goals, nor should they become such. Deming tells a story from a company which, he confesses, is one of his clients! In March 1989 he saw some figures from that company's Committee on Health and Safety. This was a list of time lost from accidents per division for both 1988 *and 1989!* Not surprisingly, he couldn't understand where the 1989 figures came from. He eventually discovered that they were just the 1988 figures reduced by 15%. That was the goal. But how was this to be achieved? Of course, people need figures—but not figures in isolation; they also need the education, training, systems, and methods to make it reasonable for the figures to be attained. After all, recalling again the important set of "Questions and Pronouncements" from Lloyd Nelson (*Out of the Crisis*, page 20), if the improvement can be achieved without a plan, why hasn't it been done already? ("You must have been goofing off!" says Deming.)

As John Dowd recently pointed out to me, it can be helpful to think of figures in three categories. First, there are facts of life, such as that referred to later in this chapter in Deming's list of "Faulty Practice" versus "Better Practice": "if we don't decrease faulty product to 5% by the end of this year, we shall no longer be here." Obviously, this only qualifies for a "fact of life" if such is *known* to be true—which is rather different from being somebody's opinion or brainchild of a good figure at which to aim. Secondly, there are figures needed for planning: predictions and budgets. These should be neither arbitrary nor targets. Predictions need knowledge—and they are for guidance and help, not for judgment and blame. And third, there are the arbitrary numerical targets of which Point 11 wants to rid us. The first two categories, used in appropriate ways, are helpful and indeed necessary. The third is not.

In his summary of the 14 Points on pages 23–24 of *Out of the Crisis*, Deming refers explicitly to Management by Objective (MBO). Recently he referred, with some despair apparent in his voice, to a report in the *Washington Post* that the President of the United States had ordered a hold-back on resources from organisations not using MBO. "What could be worse?" he asked. And then, more forgivingly (at least as far as the President was concerned):

> *"How would he know?*
> *His economic advisors didn't tell him."*

In similar vein to MBO, Brian Joiner often talks of MBR —Management by Results, or MBC—Management by Control. If we accept Point 9 (Break Down Barriers), or think of production viewed as a (horizontal) system rather than in terms of the traditional (vertical) management hierarchy (see Chapter 8), then MBR is illogical and irrelevant. If a target or quota is beyond the capability of the system to reach, it is often still reachable—but only by "beating the system," the effect of which is invariably to cause added trouble to other parts of the system, i.e. to other departments and to other people.

Even outside a Deming environment, the real effects of MBO, MBR, or MBC, whichever you like to call it (the terms are all much of a muchness), can be crazy. There are several illustrations in *Out of the Crisis* (particularly, of course, in the discussion of Point 11) of ingenuity shown in beating the system, and of the harm resulting from management's insistence on managing by quotas. Even *reasonable* quotas, i.e. those which are within the capability of the system, are harmful. A consequence of reasonable quotas is presented on page 71 of *Out of the Crisis*:

> "One will see any day, in hundreds of factories, men and women standing around the last hour or two of the day, waiting for the whistle to blow. They have completed their quotas for the day; they may do no more work, and they can not go home. Is this good for the competitive position of American industry? These people are unhappy doing nothing. They would rather work."

However, rather than repeating illustrations from *Out of the Crisis*, let us consider some further examples that are not in the book.

There was the nuclear plant which was averaging 12 serious accidents per year. Top management ruled that the rate should be halved (the aim should be to have 6 serious accidents per year!). What did the local management do? They brought in outsiders to do the dangerous jobs. Accidents in which no regular employees were injured did not have to be included in the records. The outsiders were not so well-trained as the regular employees, but their accidents didn't get into the figures. At one site there were two deaths—but no accidents were reported.

Another company had a great idea: the employees would get a $300 bonus if there were no accidents. As you might expect, no accidents were recorded. There were people hobbling around on crutches or with their arms in slings. But the official figures proclaimed that there had been no accidents.

In both cases, accident figures *appeared* to be improving miraculously. I wonder who believed them? In truth, suppression of the real figures was harmful, since problems, affecting life and limb, could now be ignored and indeed worsen while management were congratulating themselves on their success.

And, moving not far away from the subject of accidents, some airlines apparently offer bonuses to mechanics if they get their maintenance and repair work done on time. Would you like to fly on those aircraft? You probably do.

At a seminar in Newport Beach, California, Professor John Whitney told of a time when he was manager of a store. He was encouraged to reduce the shrink-rate from 4% to 1%. The "shrink-rate" is the proportion of goods which get onto the shelves but leave the store without being paid for. The encouragement was a 30% bonus if he made the target, a probationary period if he didn't, or being fired if he was way off target. Not surprisingly, John decided that his top priority was to reduce that shrink-rate. He figured out over 50 things that he could do, including hiring extra security personnel (to cut down on pilfering) and stocking much less of items (e.g. some fruits) which had relatively high spoilage rates. Costs went up and customer satisfaction went down, but John made the target and got his bonus.

Brian Joiner tells of a company that appeared on a list of the best-managed companies in America. However, even best-managed companies have good years and bad years. Toward the end of one of the leaner years, management really put the pressure on the salesmen to get out and sell as they had never sold before. They also were given a mixture of threats and promises as encouragement. They did it. They had a spectacular month, the best ever. The champagne flowed.

But then the trouble started. The Manufacturing Department had been given no warning of the suddenly-increased demands to be made on them. They did not have the capacity, even though they did put in considerable overtime. The salesmen's promised delivery dates came and went. Suppliers were thrown into chaos on receipt of substantially-increased orders which were needed instantaneously.

The company had an automatic forecasting system. Responding to the apparent huge surge in demand, the forecasting system foresaw a wonderful future. The top executives decided to buy new plant to cope with all the new business.

But what happened to sales in the following month? Of course, they slumped. The salesmen's mammoth efforts had exhausted the market. Customers had been hurried into contracts which they would normally have been completing during the next few weeks. The stockpile of leads had been used up. Many salespeople working in the quota system find it pays to indulge in "sandbagging." Now the sandbags had split.

Manufacturing suddenly found themselves with excess capacity. The suppliers, having got tooled up for increased business, were suddenly told to stop—unused inventory had rocketed.

Yes, the salesmen had beaten the system in the final month of the financial year—but at what cost? Recall the statement from Chapter 8:

> "The performance of any component sub-process is to be evaluated in terms of its contribution to the aim of the system, not for its individual production or profit, nor for any other competitive measure."

Deming tells of a related occurrence in a store in Minneapolis which came up with an exciting new scheme of sales incentives. There were great rewards for reaching target—so much so that the salespeople, near the end of the requisite time-period, were persuading customers to take home stuff they didn't really want, saying that, of course, they could always return it. They did. Results? Sales up, pay up, returns up, profits down. It costs a lot of money, when time and paperwork are taken into account, to process returns: the figure arrived at in this particular store was of the order of $20 per transaction.

Deming is clear in his opinion that incentive pay backfires—it looks good in the short term but is harmful when one takes the broader view. The salesman whose job is simply to sell as much as he can—and that *is* his job in most companies—will sell to a customer a more expensive machine than he really needs. Of course, it *is* good sales practice to assess accurately what the customer does need—since often the customer doesn't know. But if it becomes clear to the customer in retrospect that he has no use for the features and capacity for which he has spent his extra money, will he come back for repeat business, and will he recommend that company to his colleagues? Hardly. Or, in order to get the sale, the salesman may promise to arrange for delivery *immediately*, thus making other customers wait longer. But, of course, deliveries are not the responsibility of the Sales Department. What is the overall gain or loss? Similarly an insurance salesman, being paid on commission and bonus, may try to sell his customers more expensive policies than they need. The customer loses, the company loses (except in the short term); the salesman is likely to be the only one that wins.

A further story involves both incentive pay and an arbitrary target. A company started up an SPC programme and, at the same time, installed an incentive-pay scheme. It took about ten minutes to measure data, carry out the simple calculations and plot the point(s). An incentive was introduced for carrying out the operation in five minutes instead:

"The incentive pay plan has rendered useless the attempt to use SPC."

Incentive pay and pay for performance are included in the forthcoming list of Faulty Practices. There can be wide variation, but "the performance of anyone is governed almost entirely by the system"—remember the Experiment on Red Beads. However, as we have seen, it is not impossible to beat the system, but again we ask: at what cost? In any case:

"People can beat the system, but why should they need to?"

It is management's job to develop and improve the system so that it can provide what they want.

As an aside, Deming points out that if MBO or MBR were the secret of success then his fourth Deadly Disease (Mobility of Management) wouldn't be a disease at all—it would be healthy.

In addition to the sales-incentive story told above, Brian Joiner regales us with many further examples of the consequences of the style of management (let's refer to it as Management by Control) being criticised in Point 11. I quote here from his paper with Peter Scholtes: "Total Quality Leadership vs Management by Control":

"But there is an underside to Management by Control. Consider these examples:

• An electronics firm typically ships 30% of its production on the last day of the month. Why? In order to meet the monthly shipment quota. How? By expediting parts from around the country, by moving partially-completed instruments ahead of their place in line and, occasionally, by letting quality standards slip.

• Another firm sometimes ships incomplete instruments. A service representative then flies around the country installing the missing parts. The shipment quota for the month is met again. Profits, at least on paper, hold firm.

• A chemical firm reports that it cannot efficiently run at the mandated inventory levels, so it keeps inventories higher until June 30 and December 31 when inventories are measured. For those days, it depletes the inventories to an acceptable level, losing perhaps two days' production as a consequence.

• Many managers annually negotiate safe goals and manage to exceed them, just barely. Some managers include, on their list of negotiable goals, figures which were secretly accomplished prior to the negotiations.

• Production which exceeds the standards is stored so it can be pulled out and used another day.

• A meter reader stops at a tavern at 2 o'clock rather than exceed his expected work standard."

And, later:

"It is interesting to note that Management by Control is widely used in the Soviet Union. Typical is this story: Several years ago there was a surplus of large nails and a shortage of small ones. Why? Managers were held accountable for the tons of nails produced. Later the control was changed to the number of nails produced. This led to a shortage of large nails, since smaller nails gave higher counts."

Incidentally, I have heard Deming describe the Soviet Union as "Number 1 for goals and MBO," but I feel pretty sure that he considers the Western world to be busy catching up. He also mentions a company, described in Peters and Waterman's *In Search of Excellence*, which always shipped around 30% of the month's total output on the final day of the accounting month. (Is this the same company as that mentioned above by Brian Joiner and Peter Scholtes? I do not know.)

So much of what is being discussed here, and more, is summarised in a list of comparisons between "Faulty Practice" and "Better Practice" circulated by Deming at his presentations in recent years. He leaves no room for doubt on what he means by these terms. Faulty Practice is reactive and is accomplished merely by skills—it involves no knowledge, no theory, no mind:

"Most management today is just reflex action."

Another way in which Deming refers to management styles which are based on Faulty Practice is, as we have heard before, "the lazy way out." Better Practice can only come about by studying and using theory of management. Yet again, the list comparing Faulty Practice with Better Practice has grown over the years, but the following is a fairly comprehensive summary. Naturally, there is some repetition of material with which we are already familiar:

Faulty Practice	Better Practice
Reactive: skills only required, not theory of management. Mind not required.	*Theory of management required.*

Ranking of people (including salesmen), teams, plants, divisions, with reward at the top, punishment at the bottom.	A better way is to manage the whole company as a system. The function of every component, every division, under good management, is to adjust its contribution toward optimisation of the system.
Ranking creates competition between people, teams, plants, divisions: it destroys the system.	
Ranking comes from failure to understand variation: in particular, failure to understand the differences between special and common causes of variation.	Abolish ranking. Ranking is a farce. Apparent performance is actually attributable mostly to the system, not to the individual.

Faulty Practice	Better Practice
Incentive pay; pay based on performance. Performance of the individual cannot be measured, except possibly on a long-term basis. The effect of incentive pay is numbers, not quality. Result: backfire, loss. Reward for good performance may be like a reward to the weather-man for a pleasant day.	The performance of anyone is governed almost entirely by the system. Give everyone a chance to take pride in his work.
MBO, management by the numbers. ("Do it. I don't care how you do it. Just do it.") In MBO (as practised) the company's objective is parcelled out to the various components or divisions. The assumption is that if every component or division accomplishes its share, the whole company will accomplish the objective. Unfortunately, because of their interdependence, the efforts of the components do not add up.	A company will, of course, have aims: likewise, an individual will have aims. But the aim should be improvement, not to reach a number. So, a better way is to improve the system to get better results in the future. Study the theory of a system. Manage the components for optimisation toward the aim of the system. One will only get what the system will deliver. Any attempt to beat the system will cause loss.
Setting numerical goals and quotas. A numerical goal accomplishes nothing. Only the method is important, not the goal. A goal leads to distortion and faking. Of course, facts of life need to be known. Example: if we don't decrease faulty product to 5% by the end of this year, we shall no longer be here.	Work on improvement of the process. By what method? Flowcharts and the PDSA Cycle will help.

Faulty Practice	Better Practice

The so-called merit system—actually, destroyer of people. This is a form of ranking.

Under the merit system, the aim of anybody is to please the boss. The result is destruction of morale. Quality suffers.

Judging people, putting them into slots, does not help them to do a better job.

Change the system from conflict to cooperation: everybody win. Put all people on a regular system of increase in pay (see *Out of the Crisis*, top of page 118). Institute leadership of people.

Management by Results (MBR). Immediate action on any fault, defect, complaint, delay, accident, breakdown. Action based on the last data-point.

PRR (Problem Report and Resolution): tampering, making things worse.

Understand and improve the processes that produced the fault, defect, etc..

Understand variation. The distinction between common and special causes of variation is especially important in the leadership of people.

Work standards (such as quotas and time standards. They
1. double costs (there is more money spent on setting work standards and counting production than there is in actual production);
2 rob people of pride of workmanship;
3. are a barrier to improvement of output.

Study the system. Understand its capabilities, and improve them.

Provide leadership of people. Everyone is entitled to pride of workmanship. Wherever work standards have been replaced by competent leadership, quality and productivity have gone up, and people on the job are happier.

Buying materials and services at lowest bid.

Estimate the total cost of use of materials and services: purchase price *plus* predicted cost of problems in use of them, and their effect on the quality of final product or service.

Faulty Practice	Better Practice

Lack of constancy of purpose.
 	
    Short-term thinking. Emphasis on immediate results; think in the present tense: no future tense. Keep up the price of the company's stock; maintain dividends.
 	
    Pension funds, the main source of equity in the US, must by law be invested for maximum return; money churns in and out.
 	
    No number of successes in short-term problems will ensure long-term success.
	Do some long-term planning. Adopt and publish a statement of constancy of purpose.
	
	    Of course, the management must work on short-term problems as they turn up. But it is fatal to work exclusively on short-term problems, "stamping out fires".
	
	    Ask these questions: Where do we wish to be five years from now? By what method?

Failure to manage a company (or other organisation) as a system. Instead, the various components (divisions) of the organisation are individual profit-conters. Everybody loses.	Manage the organisation as a system. A system has an aim. The individual components strive for achievement of the aim of the system, not for individual profit nor for any other competitive measure. Everybody wins.

Worker training worker in succession.
 	
    It is true that he on the job knows more about most of it than anyone else, yet the practice of worker training worker takes us off to the Milky Way (see Chapter 5).
	A better way is for somebody competent to do the training.
	
	    (Note that we are talking here about training for a skill, not about education and growth.)

Delegate quality to someone, or to a group. Appoint someone as the Vice-President in charge of Quality. The result will be disappointment and frustration.	The responsibility for quality rests with the top management. This responsibility cannot be delegated.

As those who have attended a four-day seminar will know, Point 11 is one which Deming discusses with some passion. Under a heading of (in his own words): *Such Nonsense*, I can mention the following concerning MBR:

1. As we recall from Chapter 20, there are more people measuring productivity than doing anything about it;

2. Myron Tribus often describes MBR as "driving a car by the rear-view mirror";[1] and

3. How absurd it is to think of measuring the effects or results of what we have spent on education.

And watch out if the Police Department in your area runs a quota system. Deming shows a headline from a newspaper:

> "Constable Guilty of Neglect—Failed
> to Meet Quota of Arrests"

The policeman, in Toronto, was demoted because of his failure.

Perhaps Deming's most pertinent observation of all concerning Point 11 is that:

> ## *"Measurements are always
> only the tip of the iceberg."*

It's pretty dangerous to map a route through the frozen North just based on what you can see on the surface.

---

[1] Mike Dickinson has mentioned to me that a more recent version of this quotation is "driving a car by watching the white line through the rear-view mirror."

Chapter 30

POINT 12: PERMIT PRIDE OF WORKMANSHIP

Remove the barriers that rob hourly workers, and people in management, of their right to pride of workmanship. This implies, *inter alia*, abolition of the annual merit rating (appraisal of performance) and of Management by Objective. Again, the responsibility of managers, supervisors, foremen must be changed from sheer numbers to quality.

The principal thrust of Point 12 has always concerned pride of workmanship, a perception which, as we know, has been extended of late to the theme of "joy in work":

"What do you have without pride of workmanship?
Just a job,
to get some money.
There's not much joy in that."

Deming recollects instances where workers have com-
plained to him about needing to spend much of their time doing
rework. He innocently enquires of them: "But why complain? You
still get paid, don't you?" As he expected, their answers make it
clear that that is not the point. They can take no pride in their
work.

There are also several similar stories on pages 79–81 of
Out of the Crisis. Indeed, the whole of Deming's text discussing
Point 12 (*Out of the Crisis*, pages 77–85) should be studied in
detail. The reference to performance appraisal has stirred up so
many arguments over recent years that pride of workmanship,
the primary subject of this Point, has tended to receive compara-
tively little attention. Maybe that is why Deming has more recently introduced the "joy in work" concept; clearly, this is intimately connected with pride of workmanship, but it takes the
thinking a great deal further.

However, since that concept was the subject of Chapter
13, let us now follow the trend and look at some of the issues
raised by Deming with respect to performance appraisals[1] (or
merit systems, or systems of reward). An old Chinese proverb

---

[1] A complication is that the term "performance appraisal" means different things to different people. Deming's use of the term covers
schemes which involve the judgment and ranking of people, failing to
recognise that the large majority of the variation in performance comes
from the system within which people live and work rather than from the
people themselves. On the other hand, referring again to the "Modern
principles of leadership" in *Out of the Crisis*, pages 116–118, Deming
suggests: "Hold a long interview with every employee, three or four
hours, at least once a year, not for criticism, but for help and better
understanding on the part of everybody." The confusion is that some
people would also include this much more desirable activity under the
general heading of "performance appraisal." The reader is recommended to study carefully this section of *Out of the Crisis* in order to
become more aware of what is *not* included in Deming's use of the term;
this should help clarify which matters Deming really is criticising, and
why, in this current chapter.

(due, I believe, to the philosopher Lao Tze in the sixth century BC), which he is fond of quoting, will start us off:

"Reward for merit brings strife and contention."

The reader will have noticed that, not surprisingly, this topic has been touched upon many times in earlier chapters. Here are some reminders. Deming has charged the system of reward as being one of the main constraints holding us back from the Cooperation: Win-Win culture, and has indeed referred to the merit system as being a manifestation of the Win-Lose environment (Chapter 15). In Chapter 27 he asked if anyone could be blamed for maximising his own profit (i.e. suboptimisation rather than optimisation) if he gets rated that way. Item 6 in Part B of the System of Profound Knowledge (Chapter 18) refers to knowledge of the effect of the system on the performance of people. Performance appraisal was alluded to in Chapter 5 as an example of tampering with stable systems, whereas in Chapter 25 we pointed out that one of the consequences of improvement of systems is that variability in the measured performance of people will diminish.

As opposed to driving out fear, we saw in Chapter 26 that the system of reward generates and guarantees fear. The examinations given Heero Hacquebord's six-year-old daughter were a form of performance appraisal (see Chapter 10). (We shall say more on the effects of competition and grades in education in the next chapter.) The merit system is referred to as a "Faulty Practice" and as a "destroyer of people" (Chapter 29), while the loss caused by the annual rating of performance is included in the long list of important unknown and unknowable figures which Deming gives while discussing the fifth Deadly Disease (see *Out of the Crisis*, pages 121–123).

The fact that Deming regards the merit system as symptomatic of what needs to be dispensed with is seen by his referring to both the new philosophy (in Chapter 20) and the needed transformation (in Chapter 15) as "a new system of reward"; and the need for the system of reward to change came out within about two minutes of the start of one of the four-day seminars at which I have assisted:

> *"Change to the new philosophy*
> *requires abolishment of*
> *the annual rating,*
> *merit system,*
> *performance appraisal—*
> *destroyer of people by whatever name."*

Performance appraisal is seen to have smothered innovation by squashing the excitement and potential of intrinsic motivation. Remember the quote from Chapter 14:

> "People get rewarded for conforming. No wonder
> we are on the decline."

and the observation that the 98 managers out of 100 who do not have joy in their work *dare* not contribute innovation because of concern for their rating (Chapter 15).

Let us pursue this theme a little further. The aspect of bad management exposed here is the desire of some managers for their employees to simply do as they are told—that's the employee's job; that's what he is paid for. If that is the attitude, then there is neither room for change nor encouragement to contribute to change:

> *"People are unable to contribute*
> *what they'd like to contribute*
> *to their jobs;*
> *they have to concentrate on getting a good rating."*

"Joy in work,
and innovation,
become secondary to a good rating."

"People find out
what's important for merit,
and do it.
Can you blame them?"

"Stay in line;
don't miss a raise."

"The aim becomes to get a good rating,
to please the boss."

And, in consequence of the kind of stagnation involved in these quotes:

"Can anyone take pride and joy in his work·
in that situation?"

On the other hand, none of the 80 American Nobel Prize winners had to suffer performance appraisal. They had the benefits of economic independence and environments which allowed and encouraged joy in their work; they were responsible only to themselves, setting at liberty the power of intrinsic motivation which the merit system otherwise extinguishes.

Teamwork (in the sense of Joiner's "All One Team") is rendered impossible by the merit system. For who is the customer? It *should* be the real internal or external customer—the one who receives and benefits (or suffers) from the fruits of the supplier's labours. But instead it becomes the *imposed* customer, the boss, who supervises the work rather than being directly affected by it.

Closely associated with the argument is the fact that

merit rating, the third Deadly Disease, is tied in with the second Deadly Disease, the priority of short-term over long-term thinking. "Are we too hasty?" asks Deming. He considers that we are. "Can we find," he continues, "say, 12 people in all history who have achieved eminence at an early age?" He ruminates on such names as Alexander the Great, Newton, Leibnitz, Galois, and perhaps Mozart—though Mozart himself didn't think so, since late on in his regrettably short life he tore up most of his early music. No, important early achievements are rare exceptions:

"Greatness takes time."

Even in the cases of those rare exceptions, it surely takes longer than the year over which an appraisal is based! And, recalling what we have just said about the Nobel Prize winners, and our thoughts in Chapter 14 on the stifling effect of appraisal on initiative and progress, we might ponder the effect which performance rating could have had on these "early starters."

What company is not familiar with the quandary with reference to its own investment strategy? Does one invest for the long term, with consequent apparent deterioration in this year's figures, or stay looking good for the moment at the expense of letting the future look after itself? If that is the quandary at the corporate level, what chance is there of an individual taking the long-term view? Indeed, why *should* he contribute to the long-term benefit of the company if the company's short-term reaction is to penalise him for his efforts?

Another of Deming's fundamental objections to performance appraisal has always been the very large proportion of influence of the system on an individual's measured performance. He sometimes expresses this in terms of an equation like:

$$x + y + [xy] = 8,$$

or in a briefer version which incorporates the y with the $[xy]$ term. 8 is the person's measured performance, x is that person's own contribution, y is the overall system effect, and $[xy]$ represents the interaction between the system and the individual. In the Experiment on Red Beads (Chapter 6), we would have $x = [xy]$ $= 0$ and $y = 8$. That is, 100% of the measured performance comes directly from the system, with none either from the individual or through the system's interaction with him. Deming clearly believes that it is rare in practice for x to contribute any more than, say, 5% to the final figure. That may be associated with his estimate of 94% of overall problems (variation) being due to common causes and only 6% to special causes (see Chapter 4).

Also relevant is our observation in Chapter 8 that virtually any process has a far greater variety of inputs than we normally consider; yet they all have some effect on the performance and output from that process. The process of appraisal of an individual is no exception, and the individual's own personal contribution is but *one* of those inputs. Even if there were no more than one term in the above equation apart from the x (rather than the profusion actually represented by $y + [xy]$), anybody taking their first year of algebra can have no doubt that the equation cannot be solved for x. That elementary observation by itself is enough to render invalid the judgment both explicitly and implicitly produced in an appraisal of performance.

We don't (or surely shouldn't) judge a particular item of any commodity, say a sheet of steel, without considering the variation of the system within which it was produced. So why should we with people's performances?

Deming is adamant against docking people's pay for a bad item, for almost certainly the system made it:

> *"It's cruel to dock a worker for it.*
> *You'd jolly well better know it."*

While speaking of a particular female worker who was suffering under such a management stratagem (in addition to being asked to meet an impossible quota—see *Out of the Crisis*, page 72), Deming didn't rule out the principle of penalising her financially for a defective item—as long as she really and unarguably was to blame for it:

> **"If she did make it,**
> **totally independently of the system,**
> **maybe you should dock her for it."**

But how would we know? Could this really be a process with just one active input? Deming mentions a conversation with an executive of a major transportation concern in Britain who proudly told him of the way they dock a day's pay from workers who make the most errors:

> **"Can you think of anything worse?**
> **No wonder they are having strikes**
> **every Wednesday."**

—as indeed they were at the time.

Let us consider the apparently more positive side of the coin: incentive pay—pay for performance. It *sounds* great, says Deming. But think of all those examples in Chapter 29 where Management by Results went disastrously wrong. Of course, remembering Deming's admonitions, we cannot depend on examples to *establish* a theory (see Chapter 16)—but also recall that we only need one to *discredit* a theory, and we have many more than one to disprove the theory that incentive pay is the right policy. Remember also the summary dismissal of incentive pay in the list of Faulty Practices (again in Chapter 29).

Returning to *Out of the Crisis*, page 345, where there is some discussion of the activities of eight salesmen in a couple of product lines, say A and B, in the Philadelphia area, Deming now draws a diagram (omitted from his book) to clarify the arguments. Figure 51 is a copy of one of his overhead projector sketches. Clearly, Salesman 1 was underperforming. Or was he? Upon investigation, his territory was found to be Camden. "Did you ever try to sell anything in Camden?" asks Deming. Salesman 1 may well have been the hardest-working of the eight; yet under many commission systems he wouldn't have made enough to eat. There also turned out to be good reasons why Product B did not sell well in Salesman 2's territory.

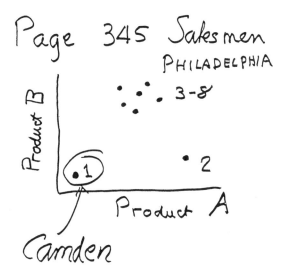

FIGURE 51
Performances of Salesmen
(Drawn by Dr. Deming)

There are two familiar lessons to repeat here. First, even when there are special causes (as in this case), don't plunge hastily into blaming the individual. And, when there are not, i.e. when representative points on a control chart lie between the 3σ limits, there is of course no logic or justification in ranking peo-

ple. Deming's clear opinion on that has already been seen in the
"Attributes of a Leader" list (Figure 50 in Chapter 25). He describes, with opprobrium, the company mentioned in Chapter 29
and in *In Search of Excellence*; in that company, an appraiser had
to take a group of five people and rank them from 1 (outstandingly good) to 5 (outstandingly bad). That company must
have had a remarkably large number of both outstandingly good
and bad people in its employ!

Deming would also like to see the education system and
environment improved to the stage where grading is no longer
used:

"The worst obsolescence in schools is ranking."

As mentioned earlier, we shall postpone most discussion on grading in schools to the next chapter. However, we include here one
intriguing point which relates to appraisal systems both in education and in industry. In grading in schools, and in performance
appraisals, Deming sees considerable evidence of the "Pygmalion Effect." From the teacher's or appraiser's viewpoint, there is
bias in expecting past good or bad performance to predict further
good or bad performance respectively—and that does affect the
detailed behaviour of the appraiser. It also affects the apprais-
ee—the employee or the student. Deming quotes George Bernard
Shaw: "Treat me like a little flower-girl: I'll be a little flower-
girl; treat me like a lady: I'll be a lady." This thought is so important that a full version of this quotation from *Pygmalion* is
also worth reproducing:

> "The difference between a lady and a flower-
> girl is not how she behaves but how she's
> treated. I shall always be a flower-girl to
> Professor Higgins because he always treats me
> as a flower-girl and always will; but I know I
> can be a lady to you because you always treat me
> as a lady and always will."

This is one of those many insights from Deming that is so easy to cast aside with scarce a thought—but, the more you think about it, the more you realise how much truth there is in what he says. And the more appalled you then become at the way we treat people both in school and in adult life.

So maybe the simple arithmetic used when discussing those two examinations which Heero Hacquebord's six-year-old daughter failed (Chapter 10) isn't valid in practice. If so, what proportion of the invalidity is accounted for by the Pygmalion Effect? It could be a great deal. And if we find that people's measured performances do not move around between some control limits as much as we might expect if we were looking at purely random variation, then again we should consider with great care how substantial the Pygmalion Effect might be.

Those that are still sceptical about this argument should study some results reported in a book entitled *Pygmalion in the Classroom*. In this book, many experiments are described in which teachers have been deliberately misled on the expectations about various children, yet in due course there has been clear evidence of the children living up to those *false* expectations:

> "20 percent of the children in a certain elementary school were reported to their teachers as showing unusual potential for intellectual growth. The names of these 20 percent of the children were drawn by means of a table of random numbers ... . Eight months later, those unusual or "magic" children showed significantly greater gains in IQ than did the remaining children who had not been singled out for the teachers' attention. The change in the teachers' expectations regarding the intellectual performance of these allegedly "special" children had led to an actual change in the intellectual performance of these randomly selected children."

In the same book we learn of children who, without fail, get high grades in one school yet, when moved to another school, find themselves at the other end of the scale. That could well be the Pygmalion Effect. There are other possible explanations. Everybody is better at some things than at others; everybody responds better to some environments than others; to some, change is traumatic because it destroys all that has been built up; to others, change is a relief because not much has been built up to destroy. Who can judge? Yet the performance appraisal ethic seems to imply that such judgment *is* quite possible, and fair, and valid. I just don't believe it.

Deming's position on performance appraisal is perceived to be one of the most controversial of all that he addresses—at least, amongst those who have not studied his thoughts in any depth. A lesser man might have dropped the issue, or at least pushed it aside, so as to make his message more palatable to the majority. Well, Deming is not that lesser man. On the contrary, he has concentrated more and more on this issue—realising that the challenge of facing this argument can result in a key breakthrough from mere on-the-surface acceptance of his philosophy to one which really gets under the skin.

Is any progress being made? It appears so. Deming points out some improvements in attitude and practice with regard to performance appraisal at Ford (see Scherkenbach's 1985 paper and *Out of the Crisis*, pages 118–119), the abandonment of compulsory performance appraisal at BOC Powertrain and other companies, and a move to abandoning a grading system at New York University. He is now receiving, on average, two letters a week telling him of the abandonment of performance appraisal systems in companies. It is still a drop in the ocean. But it seems to be a bigger drop than it was a year or two ago.

The mention of BOC reminds me of the time Deming spoke with a manager from that division of General Motors. At the start of the conversation, the manager professed total contentment with life. It didn't ring true. As the conversation continued he became more candid, and it turned out that he wasn't happy at all. In particular, he was very unhappy about having to rate the people in the division for which he was responsible. He had realised something of the illogicality and impossibility of doing that job. Nowadays, he is much happier—now he does not have to rate his people any more. The Deming message has penetrated a little further.

Here are four questions raised by delegates at four-day seminars, and the responses they received:

"What do you put in place of performance appraisal?"—

"Do you have to replace one disease by another?"

"What about people who take no pride in their work, come just for the money, etc.?"—

"They've been beaten down by the system; it's not their fault."

"How do you satisfy the need of employees for objective evaluation?"—

"It can't be done—that's why we're here. We'd better keep on studying."

"Why are performance appraisal systems proliferating throughout America?"—

"There is no substitute for ignorance."

Obviously associated with the question of performance appraisal is the matter of how people should be promoted. As an answer to the question: "How do you compensate and promote people?" Deming answered:

"I would consider anything but ranking."

Ranking people in a system is a lottery. It destroys people, the company, the country. Yes, Deming would consider anything which doesn't involve ranking. OK—so that's not a positive answer to the question. What is? We have already addressed this near the end of Chapter 25. What are our impressions of the candidates for promotion as regards their abilities as *leaders of people*, in the sense shown in the "Attributes of a Leader" list? Personal contact and personal knowledge work pretty well. Whom do you know, and who knows you? Maybe that sounds nepotistical. But, being *practical*, is there a better way?

I shall finish this chapter with a relatively recent of-, fering from Dr. Deming. I call it the "Life Diagram," and it is reproduced in Figure 52. As is so often the case when he puts pen to paper, the result is a comprehensive summary of a wealth of thought. We see time advancing across the picture, from birth on the left to death on the right. We see that at birth a baby has the whole world open to him. He is provided with the full potential of intrinsic motivation, of curiosity, of dignity, of joy, of Win-Win. What happens? The jackboots of the conventional education and management system come treading down upon him —the jackboots of grading, competition, pay for performance, MBO, and all the Faulty Practices, i.e. everything in contradiction to Points 11 and 12 and much more of the Deming philosophy. As a result, that potential available from birth gets steadily whittled away and crushed, to be replaced by the products of the competitive, selfish, extrinsically-motivated society:

"They squeeze out from an individual,
over his lifetime,
his innate intrinsic motivation, self-esteem, dignity,
and build into him
fear, self-defence, extrinsic motivation."

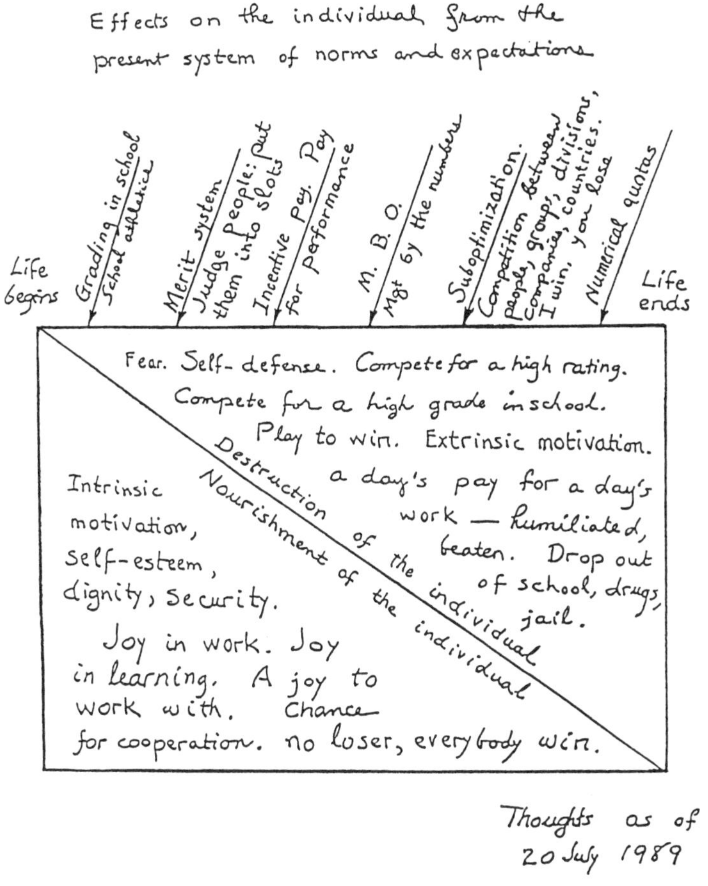

Effects on the individual from the present system of norms and expectations

Life begins / Grading in school / School athletics / Merit system / Judge people: put them into slots / Incentive pay. Pay for performance / Pay / M.B.O. / Mgt by the numbers / Suboptimization. / Competition between people, groups, divisions, companies, countries. I win, you lose / Numerical quotas / Life ends

Fear. Self-defense. Compete for a high rating. Compete for a high grade in school. Play to win. Extrinsic motivation. a day's pay for a day's work — humiliated, beaten. Drop out of school, drugs, jail.

Intrinsic motivation, Self-esteem, dignity, security. Joy in work. Joy in learning. A joy to work with. Chance for cooperation. no loser, everybody win.

Destruction / Nourishment of the individual / of the individual

Thoughts as of 20 July 1989

FIGURE 52
The Life Diagram

(Drawn by Dr. Deming)

It's a depressing picture. But doesn't it have rather more than just a ring of truth about it?

However, the purpose of the picture is not to depress. As with all Deming's teachings, its purpose is to enable us to learn what needs to happen in order to halt the decline and turn the ship around. It is the Faulty Practices bearing down upon life across the top of the picture that inexorably increase the destruction and reduce nourishment of the individual. Without them, nourishment would continue throughout life, and the destruction-half of the picture would not be there. The message is clear. The Faulty Practices must be reduced, weakened, and removed. With less weight pressing down from the top of the picture, the dividing-line between nourishment and destruction will be higher and to the right of the Life Diagram, squeezing destruction and enhancing nourishment:

"We must enlarge the bottom part.
Without that, we are ruined.
We can rise from the ashes like a phoenix.
But it will not happen unless you do it."

As at the beginning of the four-day seminar, Deming places responsibility on all of us that learn from him to go and apply that learning. Otherwise, he warns us that "we'll all pay."

The picture is powerful, and so is the message. Yet again, Deming calls for transformation.

Chapter 31

POINT 13: ENCOURAGE EDUCATION

**Institute a vigorous programme of education, and
encourage self-improvement for everyone.** What
an organisation needs is not just good people; it
needs people that are improving with education.
Advances in competitive position will have their
roots in knowledge.

Deming is, first and foremost, an educator. At the time of
writing, he still teaches virtually every Monday at New York
University, and that is 15 years after being created Professor
Emeritus and 25 years after normal retirement age, and it is in
spite of a workload and itinerary which would make people of
half his age blanch. However, he takes no credit for still teaching at NYU:

*"Anyone can do that.
That's part of the problem!"*

Being an educator is a two-way road; I have already mentioned in the Introduction to Part 3 how keen Deming still is to carry on learning—by that I include both learning from his own studies and learning from others. This, especially of somebody with his exceptional track-record, takes extraordinary and welcome humility:

> *"The greater somebody is,*
> *the more modest he becomes."*

And:

> *"Anybody can be useful*
> *if he knows his limitations."*

The primacy of education comes forth loud and clear as the very purpose of his four-day seminar. It is sometimes pointed out to delegates that in their usual kind of work they are *told* what to do. But here, for a change, the purpose is to *learn*. And:

> *"We're not here to learn skills;*
> *we're here for education—to learn theory."*

As we have seen, one of the Obstacles in Chapter 3 is the notion that "We installed quality control":

> *"You did NOT! You can install a piece of equipment*
> *but you cannot install knowledge."*

There is no substitute, we recall, for knowledge (Chapter 18).

Now, training for a skill is finite; as discussed in Chapter 24, it ends when performance has reached a stable state. In comparison, education is for *growth*, and that is never-ending. Education is knowledge; education is theory. Remember the call for transformation *in education* in the Summary of the System of Profound Knowledge (Chapter 18). One who concentrates on training is the "practical man"—also defined incidentally as one who practises the evils of his forefathers!—

"There are many practical men. We need them— where would we be without them?"

But (as Deming recalls was pointed out to him when he first worked at Western Electric in 1925 and 1926) all substantial advances contain much that, in the past, was called too theoretical. We noted in Chapter 29 how it is sheer nonsense to pretend, as some governments do, that one can measure the effects or results or rewards of what is spent on education. Education is priceless, beyond calculation. The future, not the past, is important, and education is vital for improving the future.

Education can, and should, commence early. Indeed it *does* commence early, from the day of birth, as a natural activity. But it is one of those aspects of nourishment of the individual which the present system of norms and expectations tends to squeeze out and destroy (recall the Life Diagram in Chapter 30). Remember how Deming calls for "fun in learning" at the start of the four-day seminar (Chapter 13). A lot of five- and six-year-olds have that fun when they start in school. Of those five- and six-year-olds, what proportion still have "fun in learning" ten years later?

"The first step in learning is curiosity."

Newborns have that curiosity. How long they retain it depends largely on the environment (the system) in which they find themselves. One who retained it for at least ten years is Patrick, the son of Tom Nolan (who is another member of the API group). The story which Deming is fond of relating concerns a chart which Patrick made, day by day, of the time of arrival of his school bus. Deming's reproduction of this plot, again taken from a roll of overhead projector transparency after one of his seminars, is shown in Figure 53. It indicates a mostly stable process, but with two clearly exceptional points when the bus was very late, indicating special causes on those days. Patrick had written on the chart the source of those special causes. On the first of the two days, the bus had had a new driver. On the second of the days, the door-closer was out of order. Patrick was, presumably without being directly aware of the fact, showing his interest in Item 1 of Part B in the System of Profound Knowledge (Chapter 18), i.e. knowledge about variation: what is the variation trying to tell us?[1]

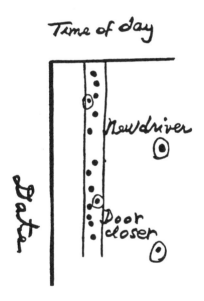

FIGURE 53
Patrick Nolan's Chart on Arrival-Times of His School Bus
(Drawn by Dr. Deming)

---

[1] The reader will note that Deming has also sketched rough control limits on the chart, showing the natural common cause variability of this process. Besides circling the two out-of-control points, he also circled the highest and lowest points between the control limits, pointing out that, quite rightly, Patrick had made no comment on these; they were just random variation, and even stable processes must have highest and lowest points!

Patrick submitted his chart as a project to his teacher. The teacher rejected it as nonsense. Little did she know that what she was rejecting was an embryo of the Shewhart control chart, particularly as it is used and annotated in Japan! Admittedly, on his chart, there were no control limits, though this was a case where the exceptional points were so clear that limits were hardly necessary. But to write identification and explanation of special causes, right there on the chart at the times the exceptional events occurred, is common practice in Japan,[2] and it helps to make the chart a living, dynamic, informative record of the behaviour of the process—it is a practice which, unfortunately, is far less common in this part of the world.

Apparently, following the rejection of Patrick's project, Tom Nolan made an appointment to go and talk with the teacher. She learned a few things. Not enough, apparently. The following year, Patrick suggested a project in which he would take a one pound weight out to various stores, put it on their scales, and observe and chart the readings. The poor teacher couldn't see the point—after all, a pound is a pound, isn't it?

Is what Patrick was doing naturally, through curiosity and intrinsic motivation, taught in our Institutes of Education, Schools of Business, etc.? We know the answer. And yet, as we also know, it is fundamental to Dr. Deming's work, and was in essence its starting-point back in those far-off days in Western Electric when he began to study Shewhart's work. Instead, even in the Business Schools, of all places:

"Half of the time is spent on learning skills.
If you want to learn skills,
go get a job, and get paid while you learn them!"

See also *Out of the Crisis*, pages 130–131. Again we recall that the Summary of the System of Profound Knowledge points out the

---

[2] See *A Japanese Control Chart*.

necessity for transformation in industry, government, and *educa-tion*. Indeed, as we saw in Chapter 18, Profound Knowledge itself should "be the backbone of courses in Schools of Business and in Departments of Statistics."

In *What is Total Quality Control? The Japanese Way,* Kaoru Ishikawa tells us that

> "Quality control begins with education and ends with education."[3]

I wonder what proportion of people who are supposedly formally involved in quality in America or Europe would place quality in the category of education rather than training. Such a one is the delegate at a four-day seminar who produced the following statement of what he considered to be top priority in influencing and encouraging the changes which are so needed:

> "Educate your customers, suppliers, and the Government about the need for constancy of pur-pose, and the tremendous costs of variation to bus-iness and to individuals; develop a better under-standing of management in Government, industry, and education."

A company should take action to aid and encourage the education of its employees at all levels. The company's own courses should no longer be restricted to training, but be broadened to include good educational content. Companies should also com-pile lists of courses available locally, prerequisites, and other information, and encourage their employees to enroll. Some com-

---

[3] A related message from Ishikawa, with similarities to Don Wheeler's statement about SPC in *A Japanese Control Chart*, is: "Quality Control is a thought-revolution in management; therefore the thought-processes of all employees must be changed. To accomplish this, education must be repeated over and over again."

panies offer to pay for "courses related to the work." Who can judge? Deming says he wouldn't draw the line anywhere. (Ford excludes basketball; Deming says he wouldn't!) The Internal Revenue Service does not help by refusing tax-relief for education "not related to the job":

"How would they know?"

How indeed![4]

Incidentally, how would the people in your company respond to such opportunities? Shouldn't this be considered as part of recruitment criteria? Recall the above wording of Point 13: "What an organisation needs is not just good people: it needs people that are improving with education." Recruitment should not just be based on past records. The recruitment process is one of the most important processes in your company. If you recruit somebody into a job who, for whatever reason, cannot carry out that job satisfactorily, whose fault is it? We have already answered that question in Chapter 24. When recruiting, look for people who are learning, and are keen to learn, and are improving, and are keen to continue improving.

Finally, we need to consider very seriously the effects of grading in education. Grades in school are equivalent to the merit system in industry: their effects are parallel:

---

[4] Kieron Dey has been kind enough to furnish me with further information on this matter. Apparently U.S. tax law decrees that, as "new" skills add value to the student, new education is not tax-deductible; however, additional education in "old" skills is job-related, is therefore of value to the company, and consequently is tax-deductible. As Kieron pointed out to me, this has led to a sorry state of affairs in some companies where those not qualified for management thus have even less incentive to study for it! The foolishness of such restricted thinking will reach tragic proportions when Business Schools get round to teaching what they should be teaching, i.e. Deming's approach. The reader will also notice total neglect here of Deming's differentiation between education and training.

"Change to the new philosophy requires abolishment of grades in school, from toddlers on up through the university."

Another version of the paragraph which we have used in Chapter 18 as a Summary of the System of Profound Knowledge makes no bones about it. It calls for transformation, particularly in education; and that transformation is to "eliminate competition, grades in school and for school athletics." The recently-added reference to athletics has been inspired by what many regard as a wholly out-of-proportion obsession with winning at sports in American colleges and universities.

Recall that one of the many reasons Deming dislikes grades is that they are invalidly and cruelly used for prediction. This came home to me years ago, long before I had even heard of Dr. Deming, when noting the lack of correlation between GCE (General Certificate of Education) Advanced Level results and the classes of degree that students were awarded just three years later.[5] And although, of course, one wouldn't claim that teaching, learning, and assessment methods in school and university are entirely the same, they are not *that* different. By and large, the students *were* in the same kinds of subjects at both stages, they *were* involved in study, they *were* involved in examinations, and so on. Even then, the Advanced Level grades were pretty poor predictors of degree classes. But GCE grades, degree classes, and all sorts of other assessments are, in effect, used in recruitment processes to assess abilities and predict performances 5, 10, 15 years ahead in jobs which often are not even remotely related in process or content to what the applicants have been doing so far. Remember Deming's scornful comment in Chapter 10 about the companies who "take the top 10% of the class."

---

[5] Even worse, experiments have been carried out in which two teachers' judgments on pupils made at the same time were compared, and resulted in correlation coefficients almost equal to zero.

398

And that is only a small part of the argument, of course. The statistical arguments against ranking people who are in a system follow easily from our work in Chapter 4, and the understanding of them is an essential feature of leadership of people (see Chapter 25). The grading of young people, from the early years, through school, and through college or university, stifles the joy of learning and smothers the intrinsic motivation to learn with which they were born. Children are put into slots. Often only so many top grades are permissible—there may be a little flexibility but, in practice, only very little. Remember the New Zealand examination system on which Deming managed to influence some change (Chapter 10). Learning, and *joy* in learning, are subjugated in the scramble to get up the list.

Actually, it isn't even clear that the supposed benefit of this system—the striving for better attainment—is effective. What happens at home when a child doesn't get good grades? In some cases he will be humiliated, depressed, chastised; in others, the reaction will be that education is "sissy" anyway, and he doesn't *want* to get good grades. In either case, the result is hardly an increase in curiosity, excitement, and fun in learning!

Often, children find themselves in a lower slot even *before* they get to school. One way that this happens is through busing—clear acknowledgement that certain schools, and thus their pupils, are inferior to others; another way is through competitive entry, which does the same thing in a different way. Who stops to think of the deep harm done to those classed as having failed before they have even started?[6] You may protest

---

[6] After reading this, Mike Dickinson wrote to me as follows concerning his experiences with Britain's "11+" examination, which used to be the mode of entry into the grammar-school system: "I was fortunate enough to have taken three 11+ examinations because I lived at the boundary of three Education Authorities. I failed the first, because I was so frightened. I passed the second, having gained from the first experience; and I won a place at the best school in the area from the third. Had I lived at another address, I would simply have been graded as a failure from the

that this argument is ignoring those who "pass" under such criteria. I really cannot believe that the pleasure, or relief, or whatever emotion results or whatever benefit ensues, is worth the harm caused to the rest. The Win-Lose balance is heavily weighted toward the Lose. On the other hand:

> *"The transformation to Cooperation: Win-Win*
> *will have profound effects on education."*

What do low grades in school mean? Surely, that the system is not treating the pupils adequately:

> *"Society pays, we all pay,*
> *—and a high price it is—*
> *for dropouts from high school.*
> *What happens to them?*
> *Drugs, crime, jail records—*
> *not all, of course, but a high proportion."*

In Chicago, the high-school dropout rate has reached approximately 45% compared with 30% some 15 years ago. How many of this huge proportion finish up in jail? Apart from all else, that's an important question in terms of sheer economics: it costs $43,000 per annum to keep someone in a Chicago jail. There is high correlation between those who spend time in jail, those not brought up in real homes, and those who get poor grades in school. Deming mentions a project which has been set up to make schools available, if needed, to play the role of students' homes for 12 hours a day. The project is funded by local industries in Chicago. How much more sensible than ranking!

Incidentally, Joyce Orsini, whom we have mentioned a couple of times already in this book, was not rated sufficiently

---

one examination, and subjected to a much inferior education system and subsequent life."

highly by a leading American university for early entry into a Doctorate program. Deming asked that her case be reconsidered. It was. She turned out to be one of his best students, and is now recognised as a Deming master. I have met her several times, and have heard her give a number of presentations on leadership and on the introductory teaching of statistics; her presentations are indeed masterly:

> *"Beginners should be taught by masters.*
> *A hack is more interesting to listen to,*
> *but teaches what is wrong."*

From what I understand, Joyce is making unprecedented progress at Fordham University in developing understanding of the Deming philosophy amongst both students and staff.

Some people have pointed out that the Japanese education system contradicts Deming's teachings on grading. He agrees. He has never said he likes everything which goes on in Japan. The situation in Japan is that there is fierce competition to enter universities, especially the more prestigious ones. But once into university there is, in effect, no further competition and, after graduation from one of these universities, a stable career is assured. A paper by Kosaku Yoshida: "Sources of Japanese Productivity: Competition and Cooperation" provides fascinating reading in this area. Dr. Yoshida obtained his doctorate under Dr. Deming at New York University in 1963.

And what of Deming at NYU? What does he do about grades? He doesn't give them. He says he has no qualification so to do. If he receives a form on which he has to enter grades, he passes all his students with the same grade:

> *"My job is to teach, not to grade.*
> *A grade doesn't help anybody."*

Refusal to grade, he says, releases the opportunity for real learn-ing. Of course, he reads his students' assignments—he learns from them, including something about how his teaching is going. Further, his students can take as long as they need to finish their main papers—why impose a deadline if he already knows what the grade will be? Sometimes they are not finished by the end of the semester: all he asks is that he be kept informed on progress —he may be able to help: [7]

"I trust them; they come through."

The quality of his students' papers is so high that he intends making them into a book (using the proceeds to aid further stu-dents).

There is, of course, another side to merit rating in edu-cation—student evaluation of courses. Deming regards such prac-tice as asinine. Again: "How would they know?" The reader may be aware of Deming's letter to the editor of *The American Stat-istician* (February 1972) on the matter.[8]

As you will imagine, Deming did not take kindly to Pres-ident Reagan's suggestion that the way to improve schools was to introduce merit pay (*Out of the Crisis*, page 103) though, as in Chapter 29, it was the President's economic advisors who were most called into question. On page 173 of *Out of the Crisis* he points out that two of the greatest teachers he has ever had, Sir Ronald Fisher and Sir Ernest Brown, would have been "rated poor teachers on every count." On the other hand, many of us must have seen teachers (including some in the quality area) holding their audiences spellbound, teaching what is wrong.

---

[7] This compares interestingly with the attitude of the grant-awarding au-thorities in Britain who in recent years have been greatly increasing the pressure for Ph.D. theses to be completed "on time" (i.e. within three years from the start of the course).
[8] This letter is reproduced in Chapter 7 of Mrs. Kilian's *The World of W. Edwards Deming*.

In a survey sent to past students, some 15 to 20 years after their graduation, Deming added two questions:

> "20. Did any teacher here affect your life?
> Yes_____   No_____
>
> 21. What was his name?
> _____ "

On studying the responses Deming found that, with rare exceptions, five particular teachers were mentioned by almost all students who had been taught by them. All other teachers were mentioned once or twice at most. Yet not one of those five would have qualified at the time as the "Greatest Teacher of the Year" (had there been such an award). The Dean of the School had made no effort to retain any of these five. He had no way of knowing that they were great. As we have seen before, "Greatness takes time."

Maybe there is an exception to the rule. Hopefully, those fortunate to be Deming's current students at NYU do judge him as great right now—as they surely will in time to come.

Chapter 32

POINT 14: TOP MANAGEMENT COMMITMENT AND ACTION

Clearly define top management's permanent commitment to ever-improving quality and productivity, and their obligation to implement all of these principles. Indeed, it is not enough that top management commit themselves for life to quality and productivity. They must know what it is that they are committed to—that is, what they must do. Create a structure in top management that will push every day on the preceding 13 Points, and take action in order to accomplish the transformation. Support is not enough: action is required.

I have exercised some discretion in my approach to the last of the 14 Points. None of the 14 Points has changed more than this one during the 1980s, but I am choosing to retain an earlier vital emphasis which is otherwise missing in more current wordings (though the content is of necessity explicitly or implicitly paramount in the other Points and in more recent main themes in the developing philosophy). That emphasis is the core role of top management in the transformation.

The change in Point 14 is clearly shown by Nancy Mann in *The Keys to Excellence*. She lists what she refers to as "the 1985 version" and "the 1986 version" of the Points. Point 14 appears in these two versions respectively as:

> "Clearly define top management's permanent commitment to quality and productivity and its obligation to implement all of these principles."[1]

and

> "Put everybody in the company to work to accomplish the transformation. The transformation is everybody's job."

On page 86 of *Out of the Crisis*, the brief wording of Point 14 doesn't mention management explicitly:

> "Take action to accomplish the transformation."

But the text immediately goes into some specific requirements for management, not referring to *top* management as such but to "management in authority."

I hope that both Dr. Deming and the readers of this book

---

[1] Another valuable observation from Kieron Dey is that many managers, whilst in early stages of understanding, look for a high-level management strategy to "go with" Deming's 14 Points. They do not realise that the 14 Points themselves, along with all that they entail, constitute the required strategy, and that what is needed in order to formulate the strategy is deep study of the 14 Points. This is totally unlike any other management strategy that they have ever seen. Kieron describes such lack of understanding as "a real issue in the United States which needs nipping in the bud."

will forgive me for, in this one instance, departing from what might currently be regarded as "the official line." However, virtually all of the content of this chapter *is* from Deming's recent teachings, though he has not referred to all of it specifically in the context of Point 14. In fact, it has been noticeable at recent four-day seminars which I have attended that, whereas Deming has discussed all the other 13 Points in detail, he has made little or no explicit reference to Point 14.

It is crucial to concentrate some attention specifically on top management. All people in management have a role to play, as Point 14 makes clear. An important element in overcoming fear of change is to get everyone "to help paddle the canoe":

> *"One of management's jobs is*
> *to manage the required change,*
> *and to involve everyone in the change."*

However, though surely there can be few top managers who are unaware that *something* is going on these days with regard to *something* called quality, most seem to be unaware that it's anything much to do with them—apart from seeing that, one way or another, others "do it":

> *"There's a lot of noise about quality.*
> *But management are washing their hands of it.*
> *Quality cannot be better than the intent.*
> *Quality cannot be delegated."*

The first reaction generally seems to be that it's all down to the shop floor: "Our quality would be all right if only the damned workers would do their job." If that is management's attitude, the workers may indeed be damned! And a President and Chief Executive Officer, Mr. G—— F——, has achieved some notoriety through often being quoted in Deming seminars. Mr. F——'s picture appeared in the February 1988 issue of the journal *Quality* above the revealing quotation:

> "Our people in the plants are responsible for their
> own product and its quality."

They are *not*, says Deming. They have nothing to do with it. They *can't*. They can only try to do their best; they can only try to do their jobs:

"The product and its quality are the responsibility of the man in the picture, the President of the company."

Mr. F—— continues:

> "We expect them to act like owners."

Then what, I wonder, is management's job supposed to be?

We have, of course, already seen a similar example in the four-point card described at the end of Chapter 21. However, in that case, if the operators fail in their responsibility "for their own product and its quality," the inspectors are supposed to save the day!

Deming also, almost without fail, refers to the boss who sees the error in the above thinking—or, rather, he thinks he does:

"The President of a company put quality in the hands of his plant managers. He was sincere—stupid, yes: but sincere."

(Remember from Chapter 1 the suggested need for confession—not

of sin, but of stupidity.) The act, says Deming, of putting quality
in the hands of the plant managers spelled doom for quality, and
the results in time became obvious and embarrassing. Why? He
goes so far as to say (and this, not surprisingly, raises some eye-
brows) that a plant manager "cannot possibly know what quality
is and, even if he did, he could do nothing about it." Like the
operators he is, in effect, helpless. One aspect of his helpless-
ness—as regards quality in the sense in which we are using the
word, as opposed to its interpretation elsewhere—is revealed in
terms with which we are familiar from Chapters 14 and 30:

> ## "He can only conform to specifications
> ## —a sure road to decline."

So what *should* top management do? Let's carry on the
progression. If the operators cannot do it, and the plant managers cannot do it, the top manager can still keep it out of his
office by appointing a Quality Manager whose task is presumably something rather more than just being in charge of the Quality Control (Inspection) Department. But what can he do? In
effect, he could be just as helpless as those others above who, in
their turn, were supposed to have been "responsible for quality."
We may recall the letter written to Lloyd Nelson, reproduced on
page 127 of *Out of the Crisis*, by someone who had just been
appointed to "the same position that you hold in your company."
The President of that company had given the new appointee "full
authority to proceed," and wished him to carry on his new job
"without bothering him." Deming comments that "it would be
difficult to convey in four lines so much misunderstanding," and
he seems to have mixed feelings as to whether the President (for
taking this attitude) or the new appointee (for taking the job)
had the greater sin (or stupidity!).
     One self-contradiction which the above letter contains is
that the writer had been appointed to the same position that

Lloyd Nelson holds in the Nashua Corporation. Lloyd's position is "Director of Statistical Methods," in effect the "Leader in Statistical Methodology," as it is referred to in Chapter 16 of *Out of the Crisis*, a similar role to that held by Bill Scherkenbach in Ford Motor Company for many years and now in the BOC Powertrain division of General Motors. Deming has recently suggested at a BDA meeting that "Leader in Profound Knowledge" might be a better title. The job is *not* that of Quality Manager; indeed:

> ## "His job is education—
> ## he is not in charge of quality."

It is a job which most certainly cannot be carried out "without bothering" the top man. Indeed, who is likely to be in the greatest need of education? Figure 61 on page 467 in *Out of the Crisis*: "Schematic Plan for Organisation for Quality and Productivity" is simple and uncompromising. The Leader in Statistical Methodology has a dotted-line link into every department and activity in the company—and an unbroken line joining him to the President of the company.

The introduction of the position of Leader of Statistical Methodology gives us, for the first time in this progression, a taste of something other than the nutritionless "instant pudding" (*Out of the Crisis*, page 126). At last the top man has stopped saying: "Get it out of my office." On the contrary, as we realised a long while ago (Chapter 14), "Quality is made in the Board Room". But:

> ## "Limitations *on quality*
> ## are also made in the Board Room."

Deming has elaborated on this in a number of ways, amongst them the following. The quality of what comes out of a company—product and service—cannot be better than the quality directed at the top—*directed at the top*, not delegated from it. A company's employees, at any level, can only produce at best the kind of product and service fashioned by the bosses:

> *"With good management,*
> *the design of product and of service*
> *will be reasonably good,*
> *helpful to the customer,*
> *suited to the market.*
> *People on the factory floor will then have a chance*
> *to do their work with pride,*
> *and to contribute to the improvement of processes."*

" ... suited to the market ... " is key. What say does the operator, the plant manager, or the Quality Manager for that matter, have regarding *what* product or service is produced. They can only do their best to do what they are told to do, and produce what they are told to produce. We recall why so many of the answers on how to obtain quality (Chapter 17) were wrong—not wrong in the sense of not being needed, but wrong because they are insufficient for the task:

> *"Of course you need good operations,*
> *but you can go out of business*
> *making without blemish*
> *a product which cannot sell."*

And not just products: Deming often refers to banks that have gone out of business even when all of their operations have been carried out "without blemish." And when plants close, he asks, is it because of poor workmanship? No: it is because what is being produced, product or service or both, does not have a market. The responsibility is management's—top management's.

411

Now, top management have taken some stick from Deming over the years. He knows it, and he admits that he may sometimes have been hard on them. We have spoken several times of "the system" within which workers work and *on* which management work. I remember the thought occurring to Deming apparently more strongly than ever before, when in discussion with British Deming Association Research Groups in July 1988, that top management themselves are also in a system and, as in any system, they too have their suppliers and customers. And they can be rather unfriendly suppliers and customers—the stockmarket, the government, the economy, the leveraged buy-out and unfriendly takeover:

> ### *"People at the top*
> ### *are handicapped in so many ways."*

True. But they are in positions of high privilege and heavy responsibility. They, even more so than everybody else in the company, have a new job. They must learn it and carry it out. Who else can do it?

Change is required, change nothing short of a real transformation:

> ### *"Management's job*
> ### *is to accomplish the change required."*

And, if it's management's job, it is most certainly *top* management's job. An official in Ford Motor Company told Deming that his job had become, in effect, manager of change. How much more so was Don Petersen's!

During the course of *Out of the Crisis* and of this book, "management's job" has been said to be many things. It is all of them. It is to learn how to change, it is to accomplish the required transformation, it is leadership of people, it is to help people, it is to enable joy in work, it is to improve systems and the working environment, it is to optimise systems rather than sub-processes (which is suboptimisation), it is to look for opportunities to widen boundaries of systems for greater service and profit, it is to focus on innovation of product and service rather than only improvement, it is to establish priorities using the Taguchi loss function, it is to aid and encourage education, it is inseparable from the welfare of the community, it is to take pride in the adoption of the new philosophy and in their new responsibilities.

    Nevertheless:

> *"A company can do well with poor management*
> *—for a while."*

I wonder, for how long?

EPILOGUE

Dr. Deming enjoys showing an advertisement for some computer software. The advertisement's headline is:

"Get in Control for Only $59.95."

He remarks:

"I think it'll take a little more than that!"

LIST OF FIGURES

BIBLIOGRAPHY

BOOKS

BOORSTIN, D. J., *The Discoverers*. Random House (1983); Penguin Books (1986).

CARLISLE, J. A. and PARKER, R. C., *Beyond Negotiation*. John Wiley & Sons (1989).

DEMING, W. E., *Elementary Principles of the Statistical Control of Quality*. Nippon Kagaku Gijutsu Renmei, Tokyo (1950); in English (1952).

DEMING, W. E., *Some Theory of Sampling*. Wiley (1950); Dover (1966).

DEMING, W. E., *Quality, Productivity, and Competitive Position*. Massachusetts Institute of Technology, Center for Advanced Engineering Study (1982).

DEMING, W. E., *Out of the Crisis*. Massachusetts Institute of Technology, Center for Advanced Engineering Study (1986); Cambridge University Press (1988).

421

DRUCKER, P., *Management, Tasks, Responsibilities, Practices.* Harper and Row (1973).

ISHIKAWA, K., *Guide to Quality Control.* Asian Productivity Organisation (1971).

ISHIKAWA, K., *What is Total Quality Control? The Japanese Way.* Prentice Hall (1985).

KILIAN, Cecelia S., *The World of W. Edwards Deming.* CEEPress Books, Washington D.C. (1988).

KOHN, A., *No Contest.* Houghton Mifflin (1986).

LATZKO, W. J., *Quality and Productivity for Bankers and Financial Managers.* American Society for Quality Control (1986).

MANN, Nancy R., *The Keys to Excellence.* Prestwick Books, Los Angeles (1985); Mercury Books, London (1989).

MORONEY, M. J., *Facts from Figures.* Pelican (1951).

ORSINI, Joyce, *Simple Rule to Reduce Total Cost of Inspection and Correction of Product in State of Chaos.* Ph.D. dissertation, School of Business Administration, New York University. (Obtainable from University Microfilms, Ann Arbor, MI.)

OUCHI, W. G., *The M-Form Society.* Addison-Wesley (1984).

PETERS, T. and WATERMAN, R. H., *In Search of Excellence.* Harper & Row (1982).

ROSENTHAL, R. and JACOBSON, Lenore, *Pygmalion in the Classroom.* Holt, Rinehart, and Winston (1968).

SCHERKENBACH, W. W., *The Deming Route to Quality and Productivity*. CEEPress Books, Washington D.C. (1986).

SCHERKENBACH, W. W., *Deming's Road to Continual Improvement*. SPC Press, Knoxville, TN (1991).

SCHOLTES, P. R., *The Team Handbook*. Joiner Associates, Madison, WI (1988).

SHAW, G. B., *Pygmalion*. E.g. *The Complete Plays of Bernard Shaw*; Paul Hamlyn (1965).

SHEWHART, W. A., *Economic Control of Quality of Manufactured Product*. van Nostrand (1931); American Society for Quality Control (1980); CEEPress Books, Washington D.C. (1986).

SHEWHART, W. A., *Statistical Method from the Viewpoint of Quality Control*. Graduate School of the Department of Agriculture, Washington (1939); Dover (1986).

WALTON, Mary, *The Deming Management Method*. Dodd, Mead & Co., New York (1986); Mercury Books, London (1989).

WALTON, Mary, *Deming Management At Work*. G. P. Putnam's Sons, East Rutherford, N. J. (1989).

WHEELER, Donald J. and CHAMBERS, David S., *Understanding Statistical Process Control*. SPC Press, Knoxville, TN (1986).

The Jerusalem Bible. Darton, Longman & Todd, London, and Doubleday, a division of Bantam, Doubleday, Dell Publishing Group, Inc., (1966).

ARTICLES AND PAPERS

DEMING, W. E., "On the Distinction between Enumerative and Analytic Surveys." *Journal of the American Statistical Association* (1953).

DEMING, W. E., "On Probability as a Basis for Action." *The American Statistician* (1975).

FINK, S. L., "Crisis and Motivation: a Theoretical Model." *Archives of Physical Medicine and Rehabilitation* (1967).

FULLER, F. T., "Eliminating Complexity from Work: Improving Productivity by Enhancing Quality." *National Productivity Review* (Autumn, 1985).

GRUBBS, F. S., "An Optimum Procedure for Setting Machines." *Journal of Quality Technology* (October, 1983).

HERZBERG, F., "One More Time. How Do You Motivate Employees?" *Harvard Business Review* (January-February 1968).

HOPPER, K., "Quality, Japan, and the U.S.: The First Chapter." *Quality Progress* (September, 1985).

HUNTER, W. G., O'NEILL, Janet K., and WALLEN, Carol, "Doing More with Less in the Public Sector." *Quality Progress* (July, 1987).

JOINER, B. L. and SCHOLTES, P. R., "Total Quality Leadership vs Management by Control." Joiner Associates, Madison, WI (1985).

NELSON, L. S., "The Shewhart Control Chart—Tests for Special Causes." *Journal of Quality Technology* (October, 1984).

SCHERKENBACH, W. W., "Performance Appraisal and Quality: Ford's New Philosophy." *Quality Progress* (April, 1985).

SCHOLTES, P. R. and HACQUEBORD, H., "A Practical Approach to Quality." Joiner Associates, Madison, WI (1987). Republished as "Beginning the Quality Transformation" (July) and "Six Strategies for Beginning the Quality Transformation" (August) in *Quality Progress* (1988).

TRIBUS, M., "Creating the Quality Service Company." SNA Mid-Winter Management Conference (1985).

TRIBUS, M., "Judging the Quality of an Organisation by Direct Observation." Massachusetts Institute of Technology, Center for Advanced Engineering Study (undated).

TRIBUS, M., "The Germ Theory of Management." Second National Conference, British Deming Association; Salisbury, England (April 1989).

TRIBUS, M. and TSUDA, Y., "Creating the Quality Company." Massachusetts Institute of Technology, Center for Advanced Engineering Study (undated).

YOSHIDA, K., "Sources of Japanese Productivity: Competition and Cooperation." St John's University, Jamaica, NY (1985).

BOOKLETS AND MANUALS

BIBBY, J., *Quotes, Damned Quotes, and ... .* Demast Books, Halifax, England (1983).

BS5750: A Positive Contribution to Better Business. British Standards Institution Quality Assurance (1987).

Continuing Process Control and Process Capability Improvement. Ford Motor Company, Dearborn, MI (1984).

Profound Knowledge, Booklet No. 5. British Deming Association, Salisbury, England (1990).

The Deming Prize. The Union of Japanese Scientists and Engineers, Tokyo, Japan (1955).

VIDEOS

A Japanese Control Chart. SPC Press, Knoxville, TN.

Continuous Improvement in Quality and Productivity. Ford Motor Company, Dearborn, MI.

Doctor's Orders. Central ITV, Birmingham, England.

If Japan Can, Why Can't We? Films Inc., Chicago, IL.

Roadmap for Change—The Deming Approach. Encyclopaedia Britannica Educational Corporation, Lake Orion, MI.

PERMISSIONS

A huge amount of material from *Out of the Crisis* and his other writings and presentations is reproduced throughout the book with the permission of Dr. Deming.

An extract from Sir John Egan's review of *Out of the Crisis* is reproduced in the Preface with the permission of the publishers of *The Director*, London.

Some data from various issues of the *Monthly Digest of Statistics* were used to help with the construction of Figure 2 (Chapter 1) with the permission of the Controller of Her Majesty's Stationery Office, London.

Extracts from *The Keys to Excellence* are reproduced in Chapters 2 and 32 with the permission of the author, Nancy R. Mann.

Two extracts from *The World of W. Edwards Deming* by Cecelia S. Kilian are reproduced in Chapter 2 with the permission of the publishers, CEEPress, Washington, D.C..

Two extracts from *The Deming Management Method*, © Mary Walton, 1986, are reproduced in Chapter 2 with the permission of the author, Mary Walton, and of the Putnam Publishing Group.

An extract from a presentation made to the Ford Motor Company in 1981 is reproduced in Chapter 2 with the permission of William E. Conway and of Edward M. Baker (representing the Ford Motor Company).

An extract from *BS5750: A Positive Contribution to Better*

Business is reproduced in Chapter 3 with the permission of the British Standards Institution (who have asked me to say that copies of Standards may be purchased from BSI Sales, Milton Keynes, England).

Numerous references to the Joiner Triangle and its components, an extract (in Chapter 3) from "A Practical Approach to Quality" by Peter R. Scholtes and Heero Hacquebord, the flow-diagram (Figure 20 in Chapter 8) of an organisation as a system, and two extracts (in Chapter 29) from "Total Quality Leadership vs Management by Control" by Brian L. Joiner and Peter R. Scholtes are all used with the permission of Joiner Associates, Madison, Wisconsin.

Some diagrams from *Continuing Process Control and Process Capability Improvement* are reproduced (with minor amendments) as Figures 6, 7, 8, and 9 in Chapter 4, and two extracts from the text of the video *Continuous Improvement in Quality and Productivity* are reproduced in Chapter 11 with the permission of Edward M. Baker of the Ford Motor Company.

An extract from a presentation by Dr. Deming to the Association Française Edwards Deming, and a description of another part of the same presentation appear in Chapters 4 and 17 respectively with the permission of Jean-Marie Gogue.

Extracts from various papers are reproduced in Chapters 4, 25, and 29, and in the Introduction to Part 2 with the permission of Dr. Myron Tribus.

Part of a letter by Mr. Robin G. Leale to *The Times*, London, January 1, 1990, is reproduced in Chapter 7 with the permission of Mr. Leale.

Figures 10 and 11 in Chapter 4, Figure 19 (with amendments) in Chapter 8, an extract in Chapter 9, and the famous "F-test example" in Chapter 21 are all reproduced from *The Deming Route to Quality and Productivity—Road Maps and Road Blocks* with the permission of the author, William W. Scherkenbach and the publishers, CEEPress, Washington, D.C..

Extracts from the London *Daily Mail* are reproduced in Chapters 7 and 14 with the permission of the publishers, Associated Newspapers, London.

Figures 21 and 22 in Chapter 8 are reproduced with the permission of F. Timothy Fuller.

A cartoon is reprinted as Figure 25 in Chapter 8 with acknowledgement from the MANS Association, The Netherlands.

Figures 30 and 31 in Chapter 9 are reproduced with the permission of Ian Graham.

The diagram of the scientific improvement process (Figure 28 in Chapter 9) and a statement about models (in Chapter 16) are reproduced with the permission of Professor George E. P. Box.

In Chapter 13, some verses are taken from *The Jerusalem Bible*, published and copyright 1966, 1967, and 1968 by Darton, Longman, and Todd Ltd., and Doubleday, a division of Bantam, Doubleday, Dell Publishing Group, Inc., and are used by permission of the publishers.

An extract from the text of the video *Doctor's Orders* is reproduced in Chapter 13, and another part of the same video is described in the Introduction to Part 5 with the permission of Malcolm Frazer of Central ITV, Birmingham, England.

Contributions to a four-day seminar in 1988 are reproduced in Chapter 15 with the permission of Thomas J. Boardman.

An extract from *The Discoverers* by Daniel J. Boorstin is reproduced in Chapter 17 by permission of the publishers, Random House, New York.

The story concerning inspection tickets in a new jacket is told in Chapter 21 with the permission of Gerald Langley.

An extract © *The Sunday Times*, December 10, 1989 is reproduced in Chapter 23 with the permission of the publishers, Times Newspapers Ltd, London.

The story concerning conditions for bereavement-leave is told in Chapter 26 with the permission of the Falk Corporation, Milwaukee, WI.

Brian Joiner's story about sales incentives is told in Chapter 29 with the permission of Films Inc., Chicago, Illinois (see Volumes III and IV of *The Deming Users Manual*).

An extract from *Pygmalion* by George Bernard Shaw is reproduced in Chapter 30 with the permission of Samuel French Inc., New York.

An extract from *Pygmalion in the Classroom* by R. Rosenthal and Lenore Jacobson is reproduced in Chapter 30 with the permission of the publishers, Holt, Rinehart, and Winston, Orlando, Florida.

An extract from *What is Quality Control? The Japanese Way* by Kaoru Ishikawa, © David J. Lu, 1985, is reproduced in Chapter 31 with the permission of the publishers, Prentice-Hall Inc., Englewood Cliffs, New Jersey.

The story concerning Patrick Nolan is told in Chapter 31 with the permission of Patrick Nolan himself.

BRITISH
DEMING
ASSOCIATION

The British Deming Association (BDA) originated from an informal meeting arranged at the University of Nottingham in May 1987 of a group of about 35 people from a wide variety of companies and national and educational institutions. Those present were mostly already somewhat familiar with the Deming philosophy, some having been to a Deming four-day seminar, others having learned of it through American parent or subsidiary companies, yet others from my introductory seminars, and some who had learned of the approach through more individual routes.

The Association came formally into existence as a not-for-profit organisation in November 1987. Dr. Deming became its Honorary Life President early in 1988. Several companies gave financial support to the Association's formation, including Albany International, Birds Eye Wall's, Hewlett-Packard, Hoechst Celanese Plastics, ICI, Iveco Ford Truck, Mars Confectionery, Sketchley Services, and Texas Instruments; other organisations involved from the beginning were NEDO (the National Economic Development Office), the Work Research Unit of ACAS (the Arbitration and Conciliation Advisory Service), and the Universities of Nottingham and Sheffield. In addition to these, other companies and organisations now represented in the Association's Research Groups include Air Products, British Rail, British Telecom, Bruhl, Cambridge LINK, Cameron Iron Works, East Midlands Electricity, the Engineering Industry Training Board, GKN Hardy Spicer, H. H. Robertson, Hydro Polymers, ICL, International Bio-Synthetics, Lloyds Bank, Lucas Engineering, Macmillan Intek, National Power, Pedigree Petfoods, Sealink, Severn-Trent Water, Shell, Silvertown Lighting, Skillchange Systems, Thorntons, Union Carbide, West Midlands Employment Service, White Arrow, Wiggins Teape, William Jackson, the Universities of Aberdeen, Bradford, Nottingham, and Salford, and several consulting companies.

The principal aims of the British Deming Association are: to promote a greater national awareness, especially among top management, of the importance of the Deming approach to businesses, organisations, the national economy, and to society in general; to help members understand and adopt the approach; to form a network for the study and exchange of information, knowledge, and experience (with theory!) between members; and to act as a link with Dr. Deming and other national and international authorities on the improvement of quality.

The Association attempts to fulfil its aims through an expanding range of activities and services including: the organisation of national and regional conferences and seminars, featuring international authorities amongst its speakers; the work of active Research Groups (see above) which explore Dr. Deming's teachings and develop practical expressions of his approach to quality improvement; the organisation of regional workshops and study groups; the provision of specialist support for User Groups of senior managers; the publication of newsletters, booklets, and technical reports; and the provision of books, reports, videos, bibliographies, and other educational materials.

It should be emphasised that Dr. Deming did not have any direct involvement with the formation of the BDA. He has never set up any kind of organisation, commercial or otherwise, to promote his teachings. And he reserved judgment at the early stages of planning for the BDA. However, he became much happier when he realised that I was not intending to set up any kind of consulting company, but rather a not-for-profit educational and research organisation. Our job is also not to "promote" Deming's teachings; it is to spread awareness of them, to further education in them, and to provide help in whatever ways we can to interested individuals and organisations who want to deepen their knowledge of the philosophy as a sure foundation for attempting to convert his thinking into practice.

Deming knows well that his approach cannot be forced down anybody's throat. Progress can only be made by those who wish to learn:

> *"Why was I in Japan in 1950?*
> *Because they asked me.*
> *I didn't go to them and say*
> *'You need me.' "*

And, perhaps with thoughts of the Deadly Diseases in mind:

> *"A psychiatrist will tell you*
> *that he can do nothing for a patient*
> *unless the patient realises that he needs help."*

From Deming's own viewpoint, the absence of any organisation, commercial or otherwise, has left him totally free to do, think, and say precisely what he believes is right and true. Had there been an organisation to support, there could have been pressure for him to make the message more commercially attractive.

But there hasn't, and we can be confident that what we read and learn from him is the real, genuine article. The non-commercial nature of the BDA similarly allows us also to stay close to the roots.

The British Deming Association's administrative office is situated at 2 Castle Street, Salisbury, Wiltshire SP1 1BB, England (telephone (0)722 412138). Enquiries are welcomed from both individuals and companies interested in furthering their knowledge and understanding of the Deming philosophy.

About the Author

Dr. Henry R. Neave has been a close colleague and friend of Dr. Deming since the mid-1980s. He has assisted at all of Deming's four-day seminars in the United Kingdom, and has participated in many other Deming seminars, conferences, and events on both sides of the Atlantic. An acknowledged authority on the Deming Philosophy, he has written this book with great encouragement from Dr. Deming, who heartily endorses it in the Foreword.

In 1987, it was Dr. Neave's inspiration which led to the formation of The British Deming Association, and he is now its Director of Research. He is also a Lecturer at the University of Nottingham, in his native England.

INDEX

Index

Index

Acknowledgement:

The author is extremely grateful to Don Wheeler for creating this index.